# Herpetological Monitoring Using a Pitfall Trapping Design in Southern California

By Robert Fisher, Drew Stokes, Carlton Rochester, Cheryl Brehme, and Stacie Hathaway, U.S. Geological Survey; and Ted Case, University of California

Chapter 5 of
Section A, Biological Science
Book 2, Collection of Environmental Data

Techniques and Methods 2–A5

U.S. Department of the Interior
U.S. Geological Survey

**U.S. Department of the Interior**
DIRK KEMPTHORNE, Secretary

**U.S. Geological Survey**
Mark D. Myers, Director

U.S. Geological Survey, Reston, Virginia: 2008

For product and ordering information:
World Wide Web: http://www.usgs.gov/pubprod
Telephone: 1-888-ASK-USGS

For more information on the USGS—the Federal source for science about the Earth, its natural and living resources, natural hazards, and the environment:
World Wide Web: http://www.usgs.gov
Telephone: 1-888-ASK-USGS

Suggested citation:
Fisher, Robert; Stokes, Drew; Rochester, Carlton; Brehme, Cheryl; Hathaway, Stacie; and Case, Ted,2008,Herpetological monitoring using a pitfall trapping design in southern California: U.S. Geological Survey Techniques and Methods 2-A5, 44 p.

# Contents

# Figures

# Tables

# Conversion Factors

Inch/Pound to SI

| Multiply | By | To obtain |
|---|---|---|
| | Length | |
| inch (in.) | 2.54 | centimeter (cm) |
| foot (ft) | 0.3048 | meter (m) |
| mile (mi) | 1.609 | kilometer (km) |
| | Volume | |
| gallon (gal) | 3.785 | liter (L) |
| | Mass | |
| pound, avoirdupois (lb) | 0.4536 | kilogram (kg) |

SI to Inch/Pound

| Multiply | By | To obtain |
|---|---|---|
| | Length | |
| centimeter (cm) | 0.3937 | inch (in.) |
| millimeter (mm) | 0.03937 | inch (in.) |
| meter (m) | 3.281 | foot (ft) |
| | Volume | |
| liter (L) | 0.2642 | gallon (gal) |
| | Mass | |
| gram (g) | 0.03527 | ounce, avoirdupois (oz) |
| kilogram (kg) | 2.205 | pound, avoirdupois (lb) |

Temperature in degrees Celsius (°C) may be converted to degrees Fahrenheit (°F) as follows:

$°F=(1.8×°C)+32$

Datum

Horizontal coordinate information is referenced to the North American Datum of 1983 (NAD 83).

# Herpetological Monitoring Using a Pitfall Trapping Design in Southern California

By Robert Fisher[1], Drew Stokes[1], Carlton Rochester[1], Cheryl Brehme[1], Stacie Hathaway[1], and Ted Case[2]

## Abstract

The steps necessary to conduct a pitfall trapping survey for small terrestrial vertebrates are presented. Descriptions of the materials needed and the methods to build trapping equipment from raw materials are discussed. Recommended data collection techniques are given along with suggested data fields. Animal specimen processing procedures, including toe- and scale-clipping, are described for lizards, snakes, frogs, and salamanders. Methods are presented for conducting vegetation surveys that can be used to classify the environment associated with each pitfall trap array. Techniques for data storage and presentation are given based on commonly use computer applications. As with any study, much consideration should be given to the study design and methods before beginning any data collection effort.

## 1.0 Introduction

Due to its mild climate, complex topography, and rich geologic and biogeographic history, southern California supports a high diversity of reptiles and amphibians (Stebbins, 1985; Fisher and Case, 1997). Most species can be rather inconspicuous, making them difficult to survey at the community level. Because of this, much of their ecology and habitat affinities are not well known. For researchers and land managers to be able to answer ecological questions and address management needs of the local herpetofauna, it is necessary to identify a cost-effective field survey technique that detects all or most of the species in a given area with minimal sampling bias. The technique that is the most accommodating to these needs is drift fences with pitfall and funnel traps.

Drift fences are barriers that act to intercept and guide small terrestrial animals into pitfall or funnel traps placed along the fences. Pitfall traps are open containers that are buried in the ground such that the tops of the containers are level with the ground. Small terrestrial animals fall into the containers as they move across the ground. Funnel traps are elongated traps that have funnels at one or both ends that allow animals to pass easily into them through the large end of the funnels. The animals, once inside the traps, have difficulty finding their way out through the small end of the funnels and are trapped.

Since initiation of large-scale pitfall trapping in coastal southern California in 1995, this technique has proven to be effective at sampling a high diversity of reptiles and amphibians, invertebrates and small mammals (Fisher and Case, U.S. Geological Survey, written commun., 2000; Case and Fisher, 2001; Laakonen and others, 2001). This report describes, in detail, all of the elements of our pitfall trapping protocol with specific descriptions of how to process reptiles (lizards, snakes, and turtles) and amphibians (frogs, toads, salamanders, and newts). The purpose of this report is to describe trap array design, materials used, sampling schedules, personnel needs, equipment requirements, logistical considerations, trap installation and operating procedures, safety precautions, specimen identification and processing, site characterization (weather and vegetation), data collection and entry, and data management and analysis.

## 2.0 Background and Justification

Monitoring biological diversity is a current priority for researchers, land managers, and resource managers. It is important to have sampling and monitoring techniques that are comprehensive, cost effective, and standardized (Dodd, 1994).

Sampling reptiles and amphibians can be difficult because of their size, behavior, and cryptic coloring. To meet this task, numerous techniques have been employed by researchers. Singly or in combination, these methods include time-constrained searches, surveys of wood debris and cover boards, quadrate searches, road "cruising", pitfall trapping, and funnel trapping (Scott, 1982; Heyer and others,1994). The latter techniques may be used in combination with drift fencing. Comparison studies of different sampling techniques have revealed that each technique has its own advantages, disadvantages, and set of sampling biases (Campbell and

[1] U.S. Geological Survey.
[2] Department of Biology, University of California at San Diego.

Christman, 1982; Vogt and Hine, 1982; Corn and Bury, 1990; Rice and others, 1994; Fair and Henke, 1997; Jorgensen and others, 1998; Ryan and others, 2002). In order to directly compare data collected over multiple sites or times, it is important to use a standardized sampling method. This method should minimize the amount of observer bias while maximizing the number of species documented. Time constrained searches, quadrat searches, and road cruising may introduce a significant amount of bias due to the different skill levels of observers. A technique that has been identified as being the most effective for trapping a wide variety of species with the least amount of observer bias in various habitats of southern California is the use of drift fences with a combination of pitfall and funnel traps (Case and Fisher, 2001; Ryan and others, 2002). In order to validate the effectiveness of the pitfall trapping technique, Case and Fisher (2001) compared results with several other survey techniques. At the same sites that pitfall traps were operated, professional herpetologists conducted timed walking transects, timed visual surveys of search plots, high intensity herpetological searching, and passive observing surveys. None of these techniques yielded results as substantial as the pitfall trapping design for determining diversity or relative abundance of the local herpetofauna (Case and Fisher, 2001).

Pitfall traps, funnel traps, and drift fencing have been used to trap herpetofauna since the 1940s. Early studies typically used traps alone or in simple linear fence arrays to collect and document herpetofauna (Imler, 1945; Fitch, 1951; Banta, 1957; Banta, 1962; Medica, 1971). The shape of the trapping array has since been adapted to meet specific habitat types and study objectives. Traditional fences with large pitfall traps are effective for sampling squamate reptiles (Imler, 1945; Gloyd, 1947; Woodbury, 1951, 1953; Nelson and Gibbons, 1972; Semlitsch and others, 1981; Enge, 2001; Ryan and others, 2002). Fences made of various materials with or without pitfall traps are effective for sampling turtles (Bennett and others, 1970; Gibbons,1970; Wygoda, 1979; Burke and others, 1998). Funnel traps placed alongside drift fences have proven effective at capturing amphibians and reptiles (Enge, 2001). In order to study population and community dynamics around ponds and ephemeral aquatic habitats, fenced arrays have been modified to encircle the water, thus catching all incoming and outgoing species (Storm and Pimental, 1954; Gibbons and Bennett, 1974; Dodd and Scott,1994). The effectiveness of terrestrial arrays has been increased by changing the shape from a linear array to an "X" or "Y" shape with trap arms protruding from a center pitfall trap (Campbell and Christman, 1982).

In addition to collecting and documenting herpetofauna, these methods have been used for studies of habitat use (Bostic, 1965; Loredo and others, 1996) and population dynamics of individual species (Pearson, 1955; Parker, 1972; Fisher and Shaffer, 1996) and communities (Storm and Pimental, 1954; Gibbons and Bennett, 1974; Dodd, 1992). More recently, the scope of these studies has increased to address current ecological issues. Arrays have been replicated

over multiple sites to look at variation within and between multiple habitats (DeGraaf and Rudis, 1990) and to study the effects of human induced impacts on the relative abundance and diversity of herpetofauna. These include effects of grazing (Jones, 1981), forestry practices (Rudolph and Dickson, 1990), mining (Ireland and others, 1994), water supplementation of desert habitats (Burkett and Thompson, 1994), and habitat fragmentation (McCoy and Mushinsky, 1994).

To date, many studies are limited to use of one type of trap, a short time frame, and (or) are conducted over a localized geographical region. We have used a standardized array of pitfall traps, funnel traps, and drift fencing to perform long-term research over a wide geographic area with replicates within and between site localities, habitats, and environments (Fisher and Case, 2000; Fisher and Case, U.S. Geological Survey, written commun., 2000; Rochester and others, U.S. Geological Survey, written commun., 2001). The large scope of the trapping effort over southern California has enabled us to evaluate not only the effect of local landscape features on herpetofauna assemblages, but also the effects of larger area geographical variables such as latitude, altitude, and climate (Fisher and Case, 2000; Fisher and Case, U.S. Geological Survey, written commun., 2000). We have used the data to study autecology of sensitive species, effects of habitat fragmentation and introduced species on native wildlife, regional patterns of herpetofaunal diversity, and historic versus current species distributions (Case and Fisher, 2001; Laakkonen and others, 2001; Fisher and others, 2002). By collecting tissue samples from animals captured in the traps, we have enabled researchers to study phylogeny and population genetics of individual species (Maldonado and others, 2001; Jockusch and Wake, 2002; Richmond and Reeder, 2002; and Mahoney and others, 2003). The purpose of this report is to provide valuable data for both the theoretical and applied sciences as well as for conservation planning.

# 3.0 Array Design

Array designs can be variable. Compared with four-armed arrays, three-armed arrays yield comparable results, (Heyer and others, 1994), use less material, and take less time to construct. Our array design consists of three 15-m arms of drift fence with seven pitfall traps and three funnel traps (fig. 1). One pitfall trap is placed in the center of the array with each of the three arms of drift fence extending outward from the center trap. The angle formed by the array arms around the center trap should be approximately 120 degrees. Pitfall traps are placed in the middle and at the end of each arm of fencing. One funnel trap is placed along each arm approximately halfway between the middle and end pitfall traps on the right side of each arm when looking from the center trap toward the end trap. The funnel traps are placed consistently on the same side of the fence at an array to ensure that funnel trapping results are comparable among arrays.

A.

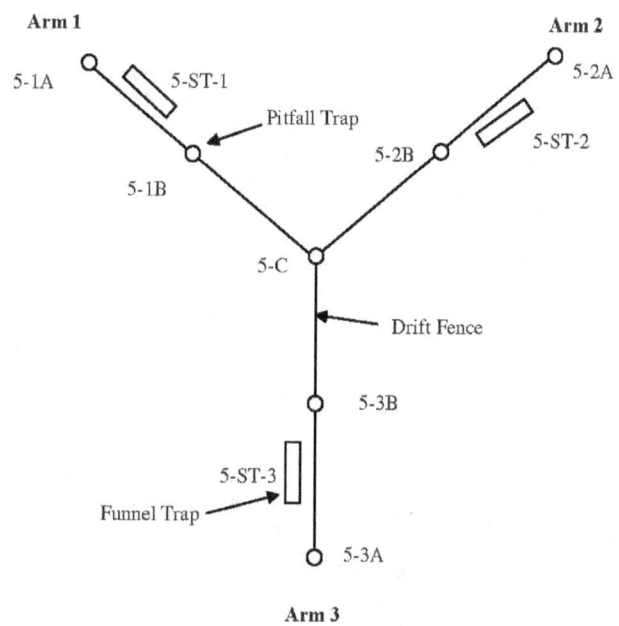

Arm 3

B.

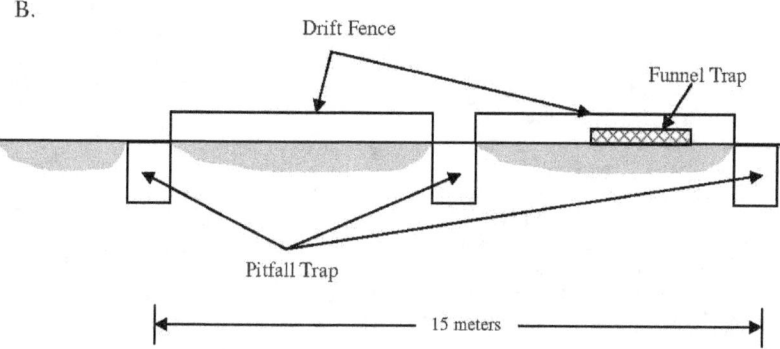

Figure 1. Pitfall array (A) overhead view and (B) side view design.

## 3.1 Trap Labeling/ Numbering

For identification purposes, each array at a given study site is assigned a number. Each array arm is assigned a number 1, 2, or 3 in a clockwise direction beginning with the arm arbitrarily designated as arm number 1 (usually the arm first encountered on the trail is assigned the number 1). The pitfall containers are labeled A, B, and C for outer, middle, and center buckets, respectively. All of the traps making up an array are first identified by the number of the array followed by the number of the arm along which the container lies. For example, the middle container at array five of arm three would be identified as 5-3B. The center container of array five would be identified as simply 5-C. The funnel traps, also called "snake traps" or "ST", are identified by the arm number along which they lie, 1–3. For example, the funnel trap located along arm two of array five would be identified as 5-ST2 (fig. 1).

# 4.0 Array Materials and Trap Construction

Below is a complete description of materials needed for the pitfall array design and instructions for making individual traps. Measures are presented in metric units with the exception of materials commonly available in Standard English units. Background information on traps and materials is presented also. A complete supply list for array construction and operation is provided in appendix 1.

## 4.1 Drift Fencing

A variety of different materials can serve as effective drift fencing and should be chosen to suit the substrate(s) and weather conditions within the study area. For instance, a porous material should be used in areas with high winds to prevent winds from tearing out fences. Other materials that have been used in various studies include clear roll plastic, silt fencing, wooden boards with bronze window screen, aluminum flashing, hardware cloth, and galvanized metal (Milstead, 1953; Storm and Pimental, 1954; Pearson, 1955; Gibbons and Semlitsch, 1981; Campbell and Christman, 1982; Bury and Corn, 1987; Murphy, 1993; Enge, 1997; Jorgensen and others, 1998; Zug and others, 2001). We use 30-cm tall nylon shade cloth. The length of drift fence is 15 m per array arm (7.5 m between each pitfall trap). Wooden stakes (1- × 2- × 24-in.) are used to secure the drift fence upright. The drift fence is secured to the stakes by using heavy-duty staples and a heavy-duty staple gun. Ten to 15 stakes are used per 15-m arm of fencing. See installation section in this report for procedures. Gibbons and Semlitsch (1981) discussed the advantages and limitations of the drift fence technique and considered aluminum flashing to be superior to other fencing materials in their area because it is more difficult for animals

to bypass and it does not deteriorate or rust. We use the shade cloth in rocky areas with many surface irregularities.

## 4.2 Pitfall Traps

Various containers have been used as pitfall traps such as 19-L (5-gal) plastic buckets, coffee cans, metal buckets, "lard" cans, local pottery water containers, and 208-L (55-gal) drums (Pearson, 1955; Banta, 1957, 1962; Gibbons and Bennet, 1974; Gibbons and Semlitsch, 1981; Campbell and Christman, 1982; Vogt and Hine, 1982; Yunger and others, 1992; Corn, 1994; and Zug and others, 2001). Larger containers generally capture more animals (Vogt and Hine, 1982), while smaller (more shallow) containers increase the likelihood for escape. The double-pit systems may aid in trapping larger reptiles (Friend, 1984), although smaller containers may be useful if the target is smaller and there is no need to capture the larger animals (Shoop, 1965; Gill, 1978; Douglas, 1979). Additionally, fitting a plastic collar to the top of pitfall traps may prevent some animals from escaping the container (Vogt and Hine, 1982).

Containers should be buried such that the rim is flush with the ground. The containers should have small drain holes in the bottom to minimize flooding during rain events while traps are open. We typically use white plastic 19-L (5-gal) buckets. At desert sites, we use 23-L (6-gal) buckets because they are deeper and provide increased insulation from heat. Crawford and Kurta (2000) found that black plastic buckets trapped frogs and shrews more effectively than did white plastic buckets. However, black buckets may experience high internal temperatures when exposed to intense sunlight for long periods. This could result in damage to the containers and increased trap mortality. Dodd (1992) used slanting pegboards to partially shade the black buckets in Florida.

### 4.2.1 Pitfall Trap Cover

All pitfall traps should have some form of top cover when open to shield animals from the elements. We use an inverted lid design set on top of the trap with wooden spacers (fig. 2). This design prevents most litter, sunlight, and precipitation from entering the open bucket while allowing sufficient space for small animals to enter.

To construct the top cover, cut three 6.4-cm-long pieces of 2- × 2-in. construction-grade wood at a 35-degree angle. Attach these, facing outward, around the top of the bucket lid using 1.25-in. drywall screws with SAE #10 washers. Make vertical cuts approximately every 15 cm around the outer perimeter of the lid. This will allow easier opening and closing of the traps. When the traps are open, the lid is turned over with the spacers resting on top of the bucket (fig. 2). Animals and wind occasionally remove the container lids while in the open position. When necessary, bungee cords may be used to keep the lids on. To attach the bungee cords to the pitfall traps, drill three holes, with diameters similar to the bungee cords, in

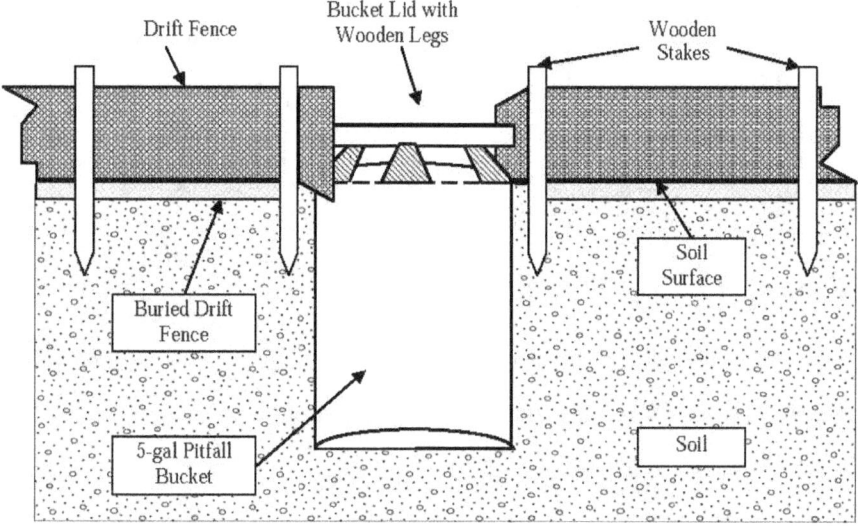

**Figure 2.** Pitfall trap design diagram (gal, gallon).

the sides of the bucket. The holes should be evenly spaced and approximately 8 cm down the container sides. Next, cut three 12-in. bungee cords in half. Feed the cord ends through the holes from the inside of the container and tie a knot at the end of the cord outside the container. Attach the cords by pulling out the hook ends and fastening onto the lid while it rests on the wooden feet. There should be enough tension in the cord that it takes considerable force to remove the lid from its resting position without unhooking the bungee cord. Wooden boards that attach to the pitfall containers using bolts and eye-sockets may also be used to keep container lids on while they are being sampled (Fellers and Pratt, 2002).

## 4.3 Equipment within Pitfall Traps

Cover should be provided within the pitfall traps for captured animals. We use two segments of differently sized PVC pipes, a 6-in. long piece of 1.5-in. diameter pipe and an 8-in. long piece of 1-in. diameter pipe. Some form of insulation such as synthetic batting or foam material should be provided in the PVC pipes if small mammals are likely to be captured. We place a section of closed foam pipe insulation within the 1.5-in. diameter piece of pipe.

Placing a wet sponge in the pitfall trap is recommended to help keep amphibians hydrated. The sponges should be wetted on a daily basis when traps are opened. The use of sponges is discontinued in southern California during the dry months, as they usually attract ants. A large number of ants can kill or seriously injure small vertebrates in the traps.

## 4.4 Funnel Traps

A variety of materials can be used to make funnel traps but 0.125-, 0.25-, or 0.33-in.-mesh hardware cloth has been the material of choice in herpetological studies (Imler, 1945; Gloyd, 1947; Fitch, 1951; and Milstead, 1953). The traps should be sturdy yet lightweight. We used 0.25-in. hardware cloth for the studies documented in this report.

To construct the funnel traps, first cut 100-ft rolls of 36-in. hardware cloth into 18-in. wide sections. For each cylinder body, roll one 36- × 18-in. section of hardware cloth along the 18-in. edge into a cylinder and fasten with hog-rings. Plastic zip-ties or cable ties can be used (fig. 3). Under field conditions, the metal hog-rings have a longer life than do the plastic zip-ties, which tend to last for only 1 year in southern California sun. The same hardware cloth is used to make the funnel ends. To make funnel ends, cut a 24-in. diameter circle of hardware cloth into four equal sections (fig. 4). Each piece is rolled into a funnel shape and serves as one end of the trap. Cut the small end of the funnel to have an opening of approximately 2-in. in diameter, which will allow animals including large snakes to enter into the cylinder. Fit two funnel ends into each cylinder with the small ends pointing inwards. Evenly cut the overhanging material in five places to the edge of the cylinder body to form flaps (fig. 5). Fold the overhanging flaps over the edge of the cylinder so that the funnels fit snugly onto the cylinder. Finally, fasten the funnels to the cylinder with medium-sized binder clips, two per funnel end (fig. 6). Trim off all sharp points and edges during the cutting process to prevent injury to captured animals and field personnel.

Ste

Ste

Ste

og rings or  lastic ca  le ties

Sea  s

Ste

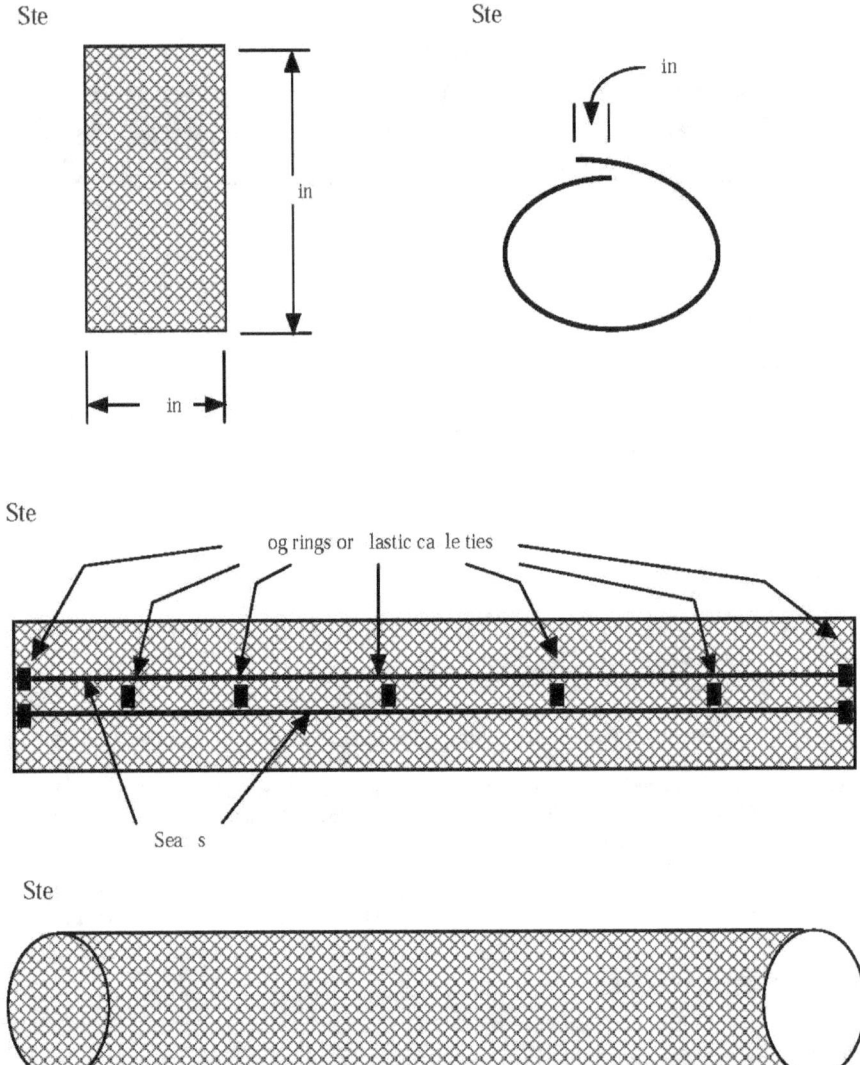

**Figure 3.**    Assembly of funnel (snake) trap body diagram (in, inch).

Ste

Ste

Ste

Ste

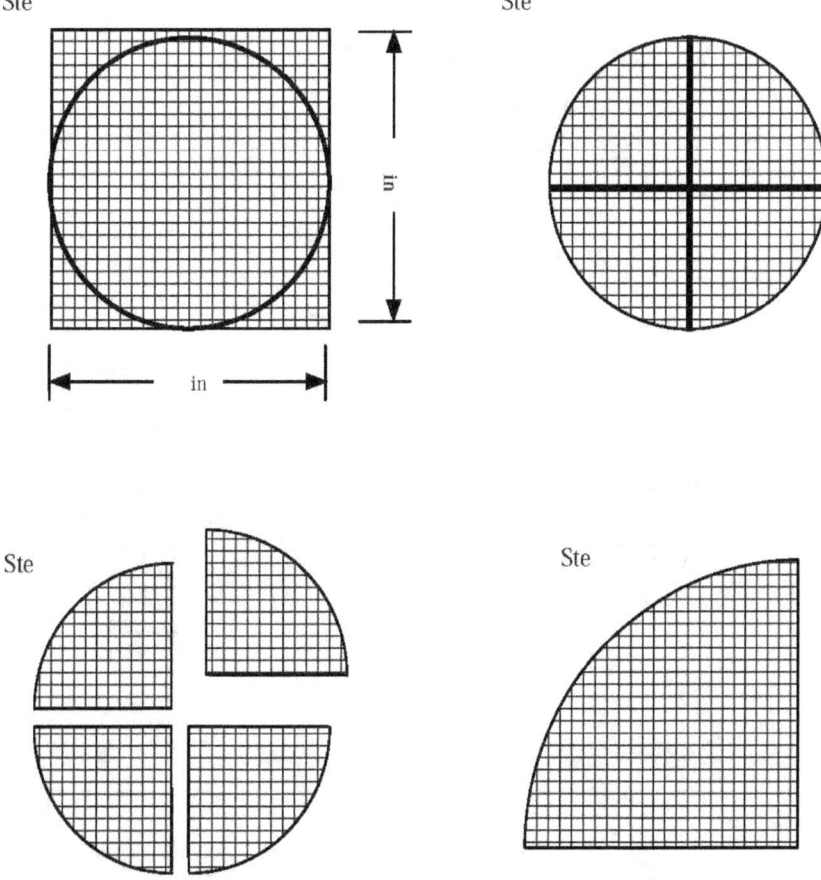

**Figure 4.**    Preparing hardware cloth for funnel (snake) trap end cone design (in, inch).

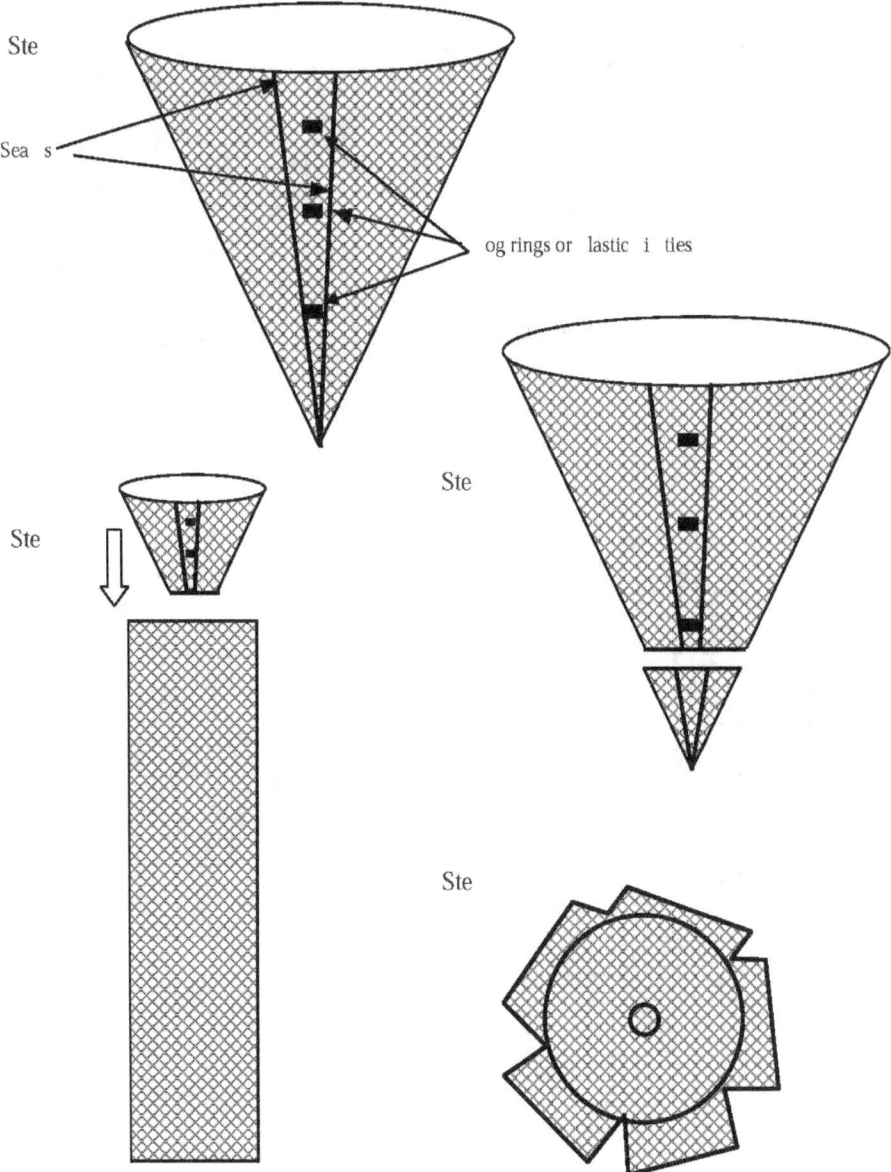

Ste

Sea  s

og rings or  lastic  i  ties

Ste

Ste

Ste

**Figure 5.**   Assembly of funnel (snake) trap and cone diagram.

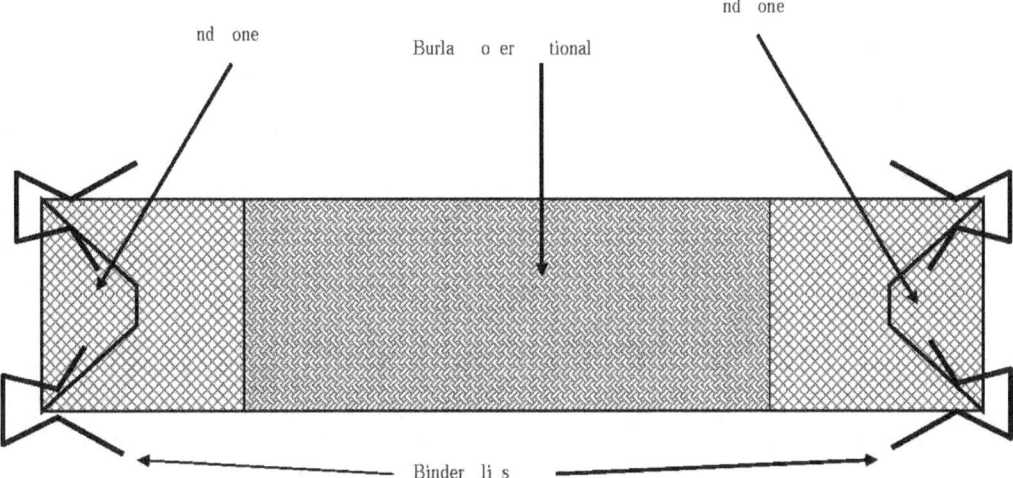

**Figure 6.** Completed funnel (snake) trap assembly diagram.

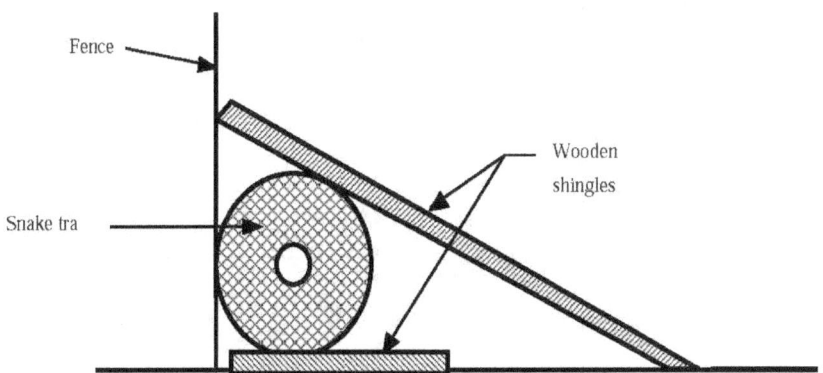

**Figure 7.** Funnel (snake) trap shade cover and position.

In the field, position the funnel traps with the seams of both the body and end cones facing upwards to prevent animals from becoming injured on any rough edges of the hardware cloth. Binder clips should be placed on the sides of the cones rather than top and bottom.

The funnel trap should be covered at all times to avoid exposure to sunlight and precipitation. We use shingle boards to cover the funnel traps (fig. 7). A shingle board is placed beneath the funnel trap also to shield it from extreme substrate temperatures. In the desert and high wind areas, 2- × 2-ft squares of 0.75-in. plywood substitute for shingle boards. A 6-in.-long piece of 1.5-in. PVC pipe with 2 in. of foam insulation is placed inside each funnel trap for additional cover.

## 4.4.1 Funnel Traps with Pitfall Trap Retreats

In areas having extreme temperatures, such as desert sites, it may be necessary to provide a more substantial secondary retreat within funnel traps. This can be done by installing a 6-gal pitfall trap under, and attached to, the funnel trap (fig. 8).

To construct this retreat, install a pitfall trap under the intended location of the funnel trap. Before rolling the hardware cloth into a cylinder, cut a hole in it using a sharpened steel pipe of appropriate diameter (fig. 9). Connect the funnel trap to the buried bucket by placing a 1.5-in. PVC "T" joint inside the funnel trap cylinder. Insert an 8- × 1.5-in.

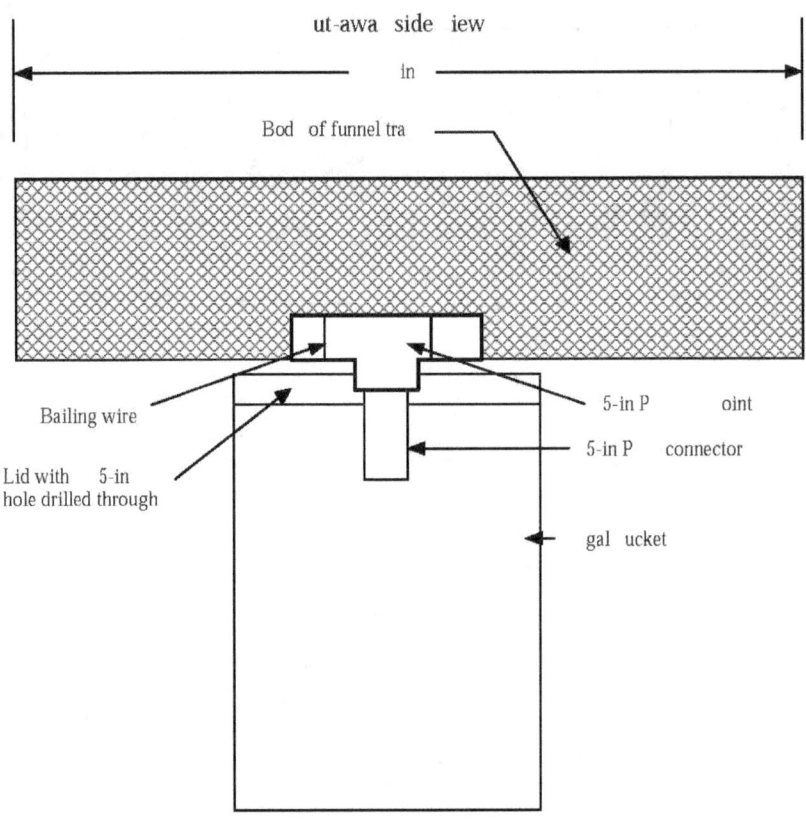

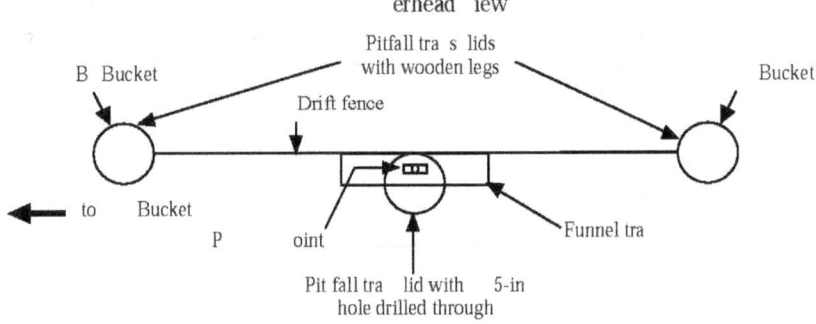

**Figure 8.** Desert funnel (snake) trap with pitfall trap retreat (in, inch; gal, gallon).

piece of PVC pipe into the trunk end of the "T" joint (fig. 8). The trunk end will extend down through the hole in the bottom of the funnel trap and into the hole cut in the top of the lid of the buried bucket. Cut the holes in the funnel trap cylinder and bucket lid just large enough to accommodate the "T" joint. Secure the "T" joint to the funnel trap cylinder with bailing wire. This design allows captured animals to seek cover in the buried bucket by way of the "T" joint pipe.

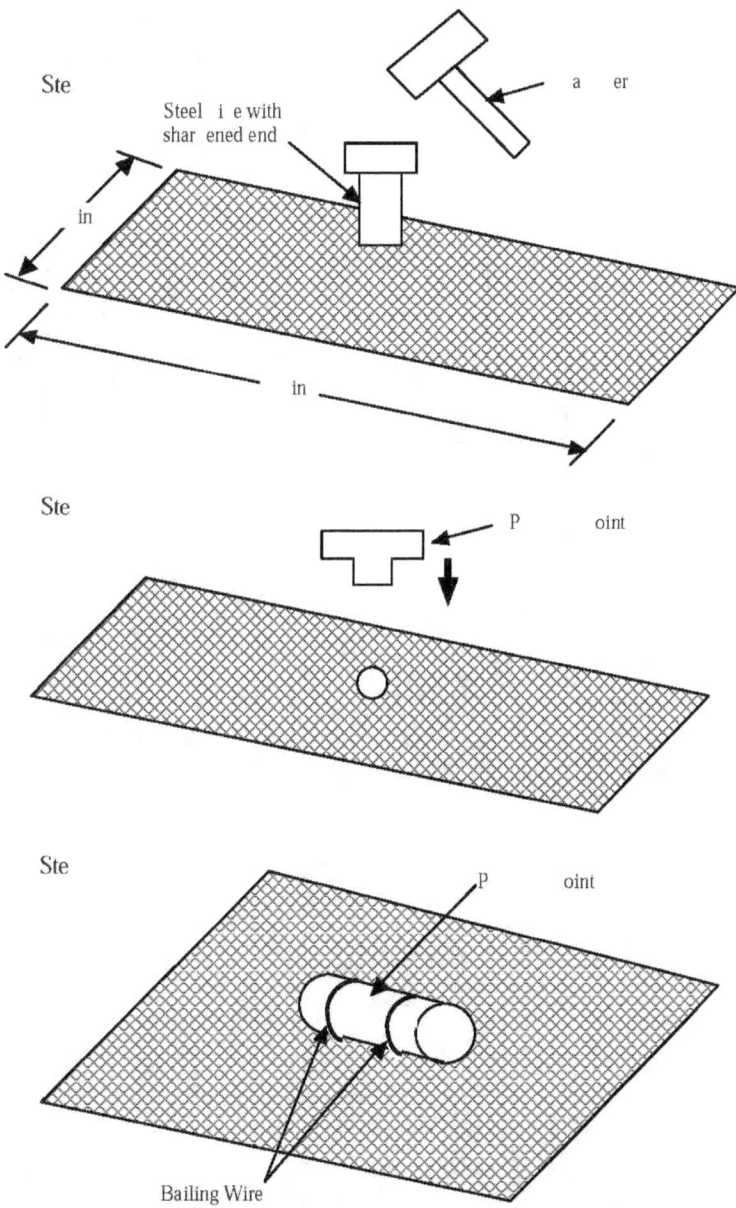

Ste

Steel i e with
shar ened end

a    er

in

in

Ste

P    oint

Ste

P    oint

Bailing Wire

**Figure 9.**    Desert funnel (snake) trap with pitfall trap retreat.

## 4.5 Weather Stations

Weather stations are designed to record ambient temperature data at a height of 1 m off the ground (fig. 10). To construct the weather station, attach an inverted 5-gal bucket to a 2- × 2-in. × 2-ft long piece of construction-grade wood using a wood bolt with a wing nut and washer. Cut away a piece from the side of the bucket that will allow for easy access to the thermometer or temperature probe (fig. 10). Insert a small wood screw into the wooden post a few inches below the top on the side of the post exposed by the section of bucket cut

away in the previous step. A thermometer may be hung on the small screw. Alternatively, a data logger device, such as HOBO (Onset Computer Corporation), may be wrapped in a plastic bag, attached to a medium-sized binder clip, and hung on the screw. Attach a piece of 2-in. diameter PVC pipe to the bottom of the wooden post using wood screws. This acts as a sleeve adapter to allow the entire station to be attached to a same sized wooden post (2- × 2- × 24-in.) buried in the ground at the selected pitfall array. If necessary, the top portion of the weather station can be moved between study sites.

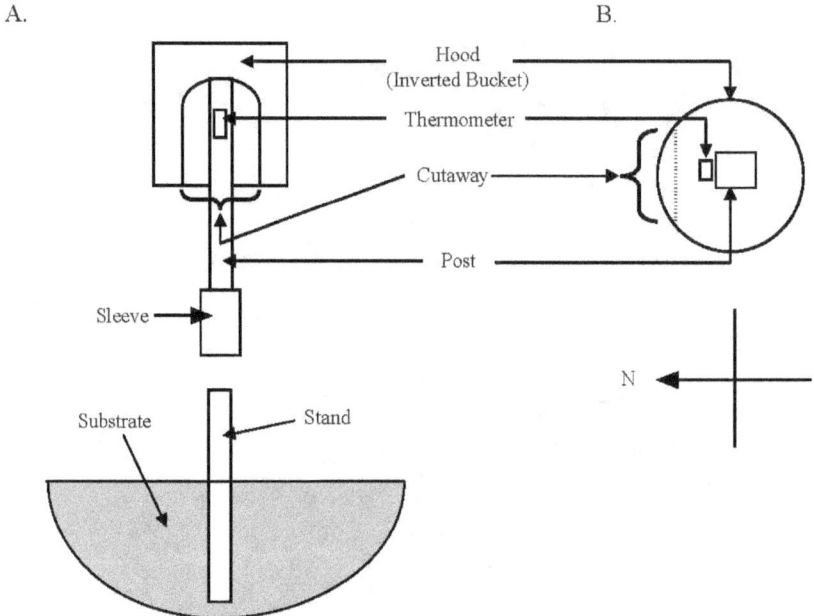

**Figure 10.**    Front view (A) and top view (B) of weather station.

# 5.0 Array Installation

## 5.1 Personnel, Equipment, and Logistics

The time it takes to install a set of arrays can vary depending on conditions of the substrate and terrain. In general, a four-person team can install a single array in about 1 hour, not including the time it takes to load/unload equipment and driving time. Considering these and other factors, usually no more than five arrays can be installed by a four-person team in 1 workday at any given study site. The equipment needed for a four-person team to install a set of pitfall trap arrays includes the following:

1. Fifty-m measuring tape to measure out array arm lengths

2. Two to three shovels

3. At least one pry-bar if working in rocky substrate or, if very rocky, a hammer drill may be necessary with a portable generator

4. (Optional) an auger with a 14-in. bit for digging holes for the pitfall traps (5- or 6-gal buckets)

5. One to two pick axes with flat blade ends to dig the trenches where the drift fencing will be placed

6. Small sledge hammer (3–5 pounds) for pounding in the wooden stakes used to support the drift fencing

7. Heavy duty staple gun and staples to attach drift fence to stakes

8. Tin snips to clip bucket lid rims

9. Scissors to cut the drift fence

10. Brush removal equipment to form trails to arrays and clear space for array arms as needed

11. At least one vehicle, with four-wheel drive if necessary, to transport the installation team and equipment to and around the study site.

Site selection involves landowner permission and (or) permits for the study in addition to any required state or federal permits for the use of this technique on the target species. Often site permits require archeological clearance that needs to be completed prior to final site selection and construction. It is better to select more sites than necessary in case some need to be abandoned during the permitting process.

## 5.2 Installation Instructions

### 5.2.1 Pitfall Arrays

During site reconnaissance, flag each of the pitfall array locations at the position of the center bucket. Measure the first array arm out 15 m from the location of the center pitfall trap. At 7.5 and 15.0 m, mark locations of pitfall traps by setting bucket lids on the ground. Measure out the remaining two arms from the center trap approximately 120 degrees apart. All three fence arms should be as straight as possible, but could bend to avoid large rocks, trees, and other barriers, as necessary.

At the marked locations of the pitfall traps, two or three crew members should begin digging separate holes using the most appropriate tools for the substrate (shovels, pick axes, pry bars). Dig the holes just large enough to accommodate the pitfall trap containers. When installing containers in substrate types that don't drain easily, it is recommended to dig the holes deeper than the depth of the containers and to place a layer of rocks below the containers before they are buried. Combined with the 0.125-in. holes drilled in the bottom of the containers, this will allow water to drain from the containers if they should become flooded during rain events. If the holes are not drilled in the bottom of the bucket then the buckets will pop out of the ground if they are closed during a rain event. If they are open during a rain event without holes, they will flood, killing most of the trapped animals.

Before burying the buckets, it is recommended that the rims of the bucket lids be clipped vertically in four or five places to make them easier to remove when opening the traps. Set the buckets into the holes with the covers on. Use dirt from the excavation to fill in the space between the outside of the bucket and the hole until the top of the bucket is flush with the ground. Having the lid on the bucket during back filling will help to maintain the shape of the bucket. Without the lid, the shape of the bucket may become deformed, making it difficult to secure the lid later. It is important to firmly compact the soil around the buried containers so that stakes put in at the edges of the buckets have a solid foundation. Next, use a pick axe to dig 3- to 5-in. deep trenches where drift fencing will be installed to connect the pitfall traps.

After all seven pitfall containers with lids are in place and the trenches are completed, install the drift fence. First, lay the appropriate length of fence along the trench. The number of stakes needed per array arm can vary, but is typically 10 to 15. For each array arm, place one stake each at the center and end containers, two stakes at the container in the middle of the array arm (one on either side), and one stake approximately every meter along the fence in between the containers. If it is necessary for the array arm to change direction or make a turn, place a stake at the inside of each turn. When the array is built on a slope, place the stakes on the down hill side of the trench for stability. Next, pound the stakes into the ground along the edge of the trench deep enough to be secure, but not so deep so that the top of the drift fence is higher than the top of the stakes. The stakes next to the containers should not be pounded into the ground until ready to attach the fencing. Pound each stake located at the edge of the container as close to the container as possible to maximize trapping efficiency, but not so close that it indents the side of the bucket.

Once the stakes are installed, set the drift fencing along the middle of the trench along the stakes. It is recommended that the fence be anchored to the end stakes (stakes at bucket edges) first. To attach the fence to an end stake, cut a 4- to 6-in. horizontal slit approximately 2 in. from the bottom of the fencing. Wrap the lower strip around the stake and staple it in place so that the top edge of the fence extends straight to and above the edge of the bucket. This flap is used to help guide animals towards the pitfall traps. Starting with the stake at the center container and keeping the fence as tight as possible, fasten the drift fence with heavy-duty staples to each stake working from the bottom of the stake to the top. Alternating sides where the stakes are stapled to the fence may be beneficial; this helps to keep the fence in place should the staples fail. Additionally, if a fence line must curve around an obstacle, the stake should be to the inside of the curve. If the fence is stapled to the inside of the curve, it often puts stress on the staples to the point of pulling out the staples and the fence falls down. When the fence has reached the container in the middle of the array arm, it should be cut such that the fence edge is extending just beyond the edge of the container and attached as described for the first end stake. Start the fence again on the other side of the middle container and fasten to the remaining stakes until the end container is reached. Install the fences of the other two array arms in this same manner until all three are in place.

Replace and pack the substrate along the drift fence such that the bottom of the fence is buried completely and all of the pitfall container rims are flush with the ground. Set the funnel traps in place along the drift fence between the pitfall containers in the middle and the outer end of the array arms. The funnel traps are placed along the right side of each arm when viewed from the center bucket (fig. 1). Close all bucket lids and remove funnel trap ends until sampling begins.

### 5.2.2 Weather Stations

Choose a location for the weather station that best represents the study site. Pound a wooden post (2- × 2 × 24-in.) into the ground at the chosen location. Leave 1 ft of

the stand exposed above the substrate (fig. 10) so that when the weather station is affixed, the thermometer or data logger is approximately 1 m above the ground. Affix the weather station by sliding the exposed PVC sleeve over the top of the ground stand. Orient the station such that the cutaway portion (exposing thermometer or data logger) is facing north in the Northern Hemisphere to prevent sunlight from reaching the thermometer or data logger. This setup allows for the collection of ambient temperature at the study site.

It may be necessary to have more than one weather station per study site, depending on the topographic diversity in the area and different microclimate zones encompassed within the study site. The weather stations can be installed at the study site when arrays are opened and then removed at the end of the sample period.

# 6.0 Array Operation

## 6.1 Survey Scheduling

The timing of surveys will vary depending on the research objectives. For instance, if surveying exclusively for amphibians, it is recommended that traps be opened after rainfall to maximize captures. However, this type of opportunistic trapping may be logistically difficult. Rain events can be unpredictable and can create access problems where use of dirt roads is required to get to the traps. Alternately, continuous trapping reveals seasonal and weather-related variations in animal activity but requires more personnel and may affect resident animals.

In our protocol, traps at any given study site are sampled for 4 consecutive days and then closed (for example, opened on Monday, sampled daily Tuesday through Friday, and closed on Friday). This is referred to as a sample period. Sample periods are scheduled every 4–5 weeks for a given site, resulting in 10–12 sample periods a year. This sampling schedule allows for target information to be collected, including seasonal activity patterns of reptiles and amphibians. Based on our analysis of species accumulation curves, it is recommended that a site be sampled 3–5 consecutive years to increase the probability of detecting rare species (fig. 11). The probability of detecting rare species also depends on the number of trapping arrays at any given site. We usually place arrays in multiple representative habitats within a site and include replicates within habitat types, as funding permits.

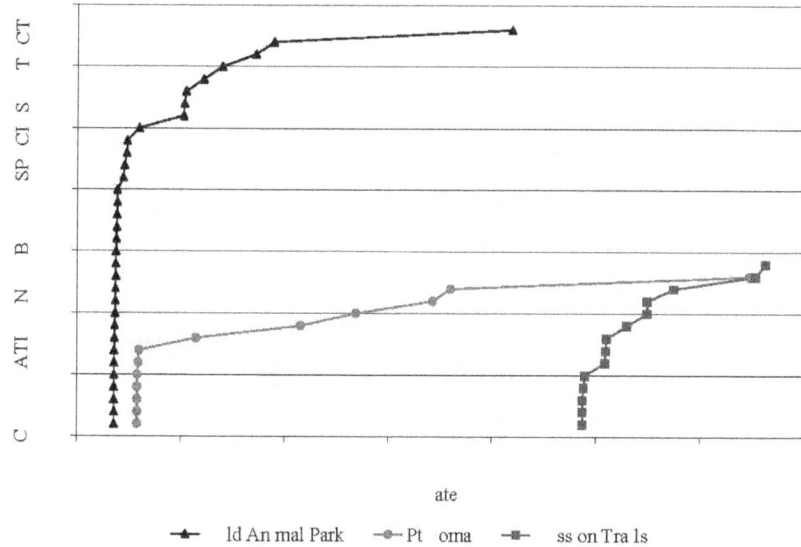

Figure 11. The cumulative number of species detected by pitfall array surveys for three study sites in San Diego County, California.

ate

▲ ld An mal Park    ● Pt oma    ■ ss on Tra ls

## 6.2 Personnel, Equipment, and Logistics

In general, a single person can operate a set of 20–30 pitfall trap arrays a day. Following the 4-day sampling schedule, a single person working full time can operate 4 sets of 20 pitfall trap arrays on a rotational basis. Some circumstances, such as long distances between arrays or personnel safety concerns, may warrant the need for a two-person team for array operation.

Each person or team operating a set of pitfall trap arrays will need a field kit that consists of a portable carrying case, such as a backpack and (or) tackle, tool, or ammunition box plus several items for processing animals and recording data. The processing equipment needed includes

1. Gloves for handling reptiles, small mammals, and (or) invertebrates.

2. A set of appropriately sized spring scales; typically 10-g, 30-g, 60-g, 100-g, and 1-kg scales will accommodate most small vertebrates

3. At least one small bag for weighing small animals (re-closable plastic bag)

4. At least one large bag for weighing larger animals (a bag or pillow case for weighing a snake)

5. Clear plastic, metric ruler

6. Cloth or plastic, 2-m measuring tape for measuring large snakes

7. Large and small forceps

8. Small scissors for toe- and scale-clipping

9. Lab marker

10. 5-mL microcentrifuge tubes containing 95-percent ethanol for storing small pieces of tissue (toe, tail, and scale clips) or ectoparasites

11. Box with dividers for storing 5-mL tubes containing 95-percent ethanol

12. 50-mL centrifuge tubes containing 95-percent ethanol for storing larger pieces of tissue

13. Large and small plastic bags for storing dead specimens that are too large for either 5- or 50-mL tubes (specimens in plastic bags should be placed immediately on ice)

14. Cooler containing dry ice or ice packs for temporary storage of dead specimens to be immediately transferred to freezer

15. Small spatula or cup for clearing out debris from within pitfall traps; a cup may also be needed to bail water out of pitfalls during rain events

16. Field identification guide to help identify captured animals

Equipment needed for recording data includes

1. Data book (small three-ring binder or hand-held computer) with appropriate data forms

2. Write-in-rain pen for recording data

3. A permanent marker for writing on plastic tissue tubes and bags

## 6.2.1 Safety Precautions

Safety precautions that should be considered when operating pitfall traps. When trapping in areas where venomous snakes occur, it is recommended that persons operating the traps wear some form of protection for their legs and ankles, such as "snake chaps." The handling of venomous snakes is described in the "Processing of Specimens" section of this report. In addition to venomous snakes, many other stinging or biting invertebrates and vertebrates are often trapped and warrant careful handling.

Precautions also should be taken in areas where the occurrence of communicable diseases, such as Hantavirus or Bubonic plague, is probable or confirmed in the rodent populations. Rodents are often caught in both pitfall and funnel traps. It is recommended that persons operating the traps wear latex and (or) thick leather gloves and some form of respiratory protection. Project leaders should check with local public health authorities before initiating fieldwork. The traps may be washed and sterilized with a diluted bleach solution as deemed necessary. It is important to ensure that the traps are rinsed thoroughly if bleach is used.

## 6.3 Survey Methods (Checking Traps)

The sample period starts by opening the pitfall and funnel traps at each array to be sampled. The PVC pipes and sponges placed into the traps should be washed in soapy water, sterilized in a diluted bleach solution (5 percent or less), and rinsed thoroughly before being brought to the site. To open pitfall traps, remove the lids from the containers. Turn over the lids and place them so that the wooden "feet" sit on the container rim. As each trap is opened, remove dirt and debris with a spatula or cup. Place one 6- × 1.5-in. diameter pipe with foam insulation, one 8- × 1-in. pipe, and one clean sponge in each pitfall trap as it is opened. Place one shingle board under each funnel trap. Place one 6- × 1.5-in. diameter PVC pipe with insulation inside each funnel trap and cover the entire trap with the shingle boards. Open funnel traps by fastening both funnel ends to the cylinder body using the binder clips.

After opening, the traps are then checked for 4 consecutive days, usually during the early morning hours. To check the pitfall traps, remove the open lids and visually inspect the inside of the containers, including inside the PVC pipes, under the sponge, and under any debris such as leaf litter and soil. During the wet season, wet the sponges daily. To check funnel traps, first visually inspect for venomous snakes, then pick up one end of the trap and inspect the inside of the trap, including inside the PVC pipe. It is usually not necessary to remove an end funnel to see inside the trap. In desert sites, the funnel traps that are connected to pitfall containers (fig. 8) should be lifted so that the inside of the pitfall container can also be inspected. All of the arrays at a study site should be checked and all animals processed and released before daytime temperatures reach levels that could result in animal mortality. If a fast moving snake (for example, *Masticophis*) or lizard (for example, *Cnemidophorus*) is captured, it may be easiest to empty the end of the trap directly into a cloth bag or pit-trap to capture the specimen for processing.

On the last day of the sample period, close the traps after checking and processing animals. To prevent animals from being captured in the traps between sample periods, remove both ends of each funnel trap and tightly attach all lids to the pitfall containers in the closed position. It may be necessary to place small rocks on top of the closed container lids to prevent them from being removed by animals or wind. Rocks that are too large will eventually break through older lids that have become brittle due to sun exposure. If periods between sampling are long or if arrays are in sites subject to vandalism, pitfall traps should be checked periodically to ensure they are intact between sample periods.

## 6.4 Field Identification

Persons operating the pitfall trap arrays should be trained to identify all of the focal species potentially trapped in the study area and should be aware of other species that might be trapped incidentally. This can be done by studying field guides, museum specimens, and (or) by training with an experienced field biologist. It is important that the person checking the traps is also able to identify the sex, relative age, and reproductive condition, when possible, of trapped animals. It may be helpful to carry field identification guides while sampling. If the field technician cannot positively identify an animal, the animal may be photographed or brought in from the field, when appropriate, for further examination.

## 6.5 Processing Specimens

All animals trapped alive are processed and released immediately. Processing captured animals involves handling, recording relevant data, and marking. Dead animals are preserved as voucher specimens. They may be collected in clean, re-closable plastic bags, 50-mL centrifuge tubes, or other appropriately sized air tight containers. The specimens can be preserved immediately in 95-percent ethanol or temporarily stored on ice in a small ice chest and transported to a freezer. The container should be appropriately labeled with site, trap, species, date, and a unique identifier such as a record identification number, if possible. Photographs of representative species can be taken also and will serve as vouchers for reports and publications.

### 6.5.1 Processing Lizards, Turtles, Frogs, Toads, Salamanders, and Newts (Limbed Animals)

To process a limbed reptile or amphibian, first record the array and trap number in which the animal was trapped. Record the species, age (juvenile/ adult), any unusual markings, deformities and (or) injuries. Record the sex and reproductive status, if possible. Measure the length of lizards, newts, or salamanders by placing a ruler on the ventral side of the animal and measuring the length from the tip of the snout to the vent (fig. 12). Frogs and toads are measured from the tip of the snout to the end of the urostyle. The length of the carapace is recorded for turtles and tortoises. All length measurements should be made to the nearest millimeter. Before recording the weight, tare the spring scale to the attached weighing bag. Place the animal into the weighing bag and record the weight as accurately as the spring scale displays.

All limbed species, except turtles and extremely small salamander species, are toe-clipped for identification purposes. A combination of toes are clipped, a maximum of one toe per foot, such that no two individuals of a species at a study site are marked with the same combination (figs. 13, 14). Toes are assigned numbers by looking at the individual from the top or dorsal side with head facing up or in the 12 o'clock position. All numbers are assigned in a clockwise direction beginning with the toes of the left front foot being assigned the numbers 1–5 (or 1–4; frogs, toads). The toes of the right front foot are assigned the numbers 10–50 (or 10–40). The toes of the right hind foot are assigned the numbers 100–500 and the toes of the left hind foot are assigned the numbers 1,000–5,000. The accelerator toes on the hind feet (toe numbers 400 and 2,000) of lizards are not clipped. Similarly, thumbs of frogs and toads (toe numbers 4 and 10) should not be clipped, as they are important for amplexus. Alternate numbering systems may be used (Heyer and others, 1994).

Some animals may be missing toes due to natural causes. In these cases, the animal can be assigned the number corresponding to its missing toes. If more than one toe is missing per foot, the highest numbered toe per foot is used by default, with all missing toes recorded in the notes. For example, if an animal was missing the number 1 and 2 toes on the front left limb, and the number 40 and 50 toes on the front right limb, it would be identified as 0052 with recapture status as unknown and the notes would read "missing toes 1, 2, 40 and 50". It is not always possible to distinguish these animals from toe-clipped individuals.

Turtles can be individually marked by making notches on the edges of the marginal scutes or by painting the shells, depending on the level of

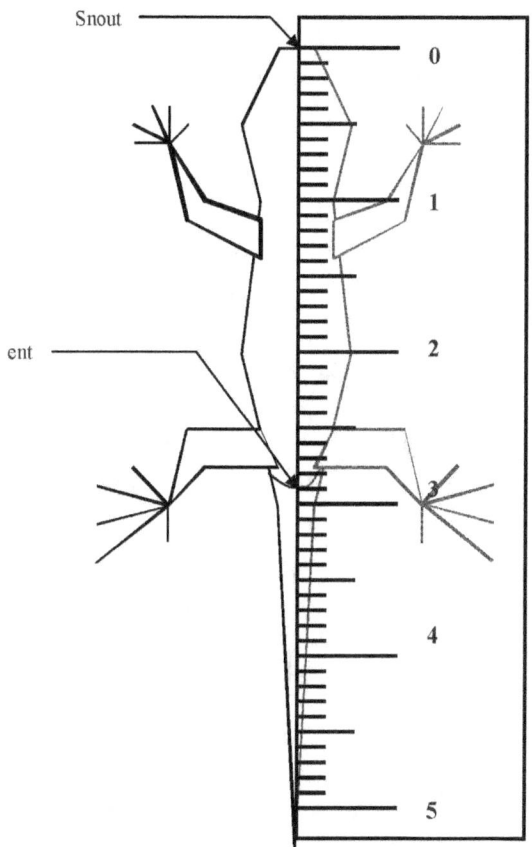

**Figure 12.**  Measure each animal from snout to vent.

permanency needed in the study (Cagle, 1939; Woodbury, 1953). If the marginal scutes are going to be marked, this should be coordinated with other researchers that are studying these species in the area or have done so in the past to share numbering schemes and data for already marked individuals. This is more important for turtles than many other reptiles due to their longevity and the possibility of multiple research efforts detecting the same individuals many years apart. Different researchers have used angle files or scissors to mark smaller turtles, but larger ones will need to be marked using a hacksaw blade or grinder (Woodbury, 1953). The numbering scheme that is used needs to be well documented in the study and associated metadata. Because of the endangered or threatened status of many turtles, we have taken measures to reduce the capture of these species. Where study sites overlap the distribution of *Gopherus agassizii*, bungee cords are used to secure the opened lids in place on top of the pitfall traps, reducing the size of the opening. Thus, we have been able to study reptile communities without the risk of harm to this species or the need for permits for it.

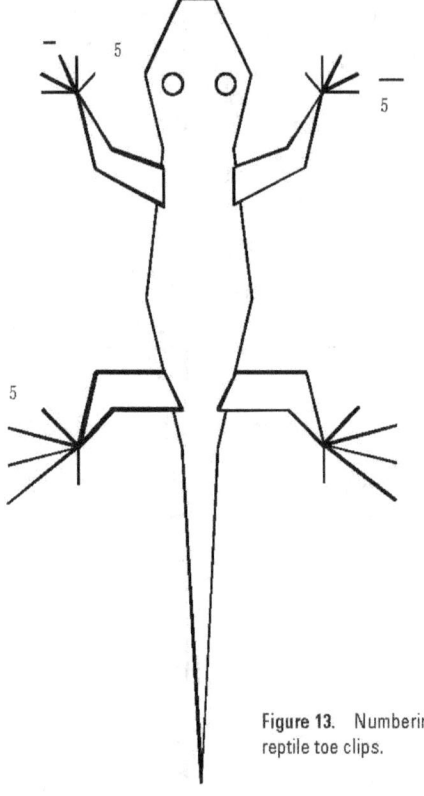

**Figure 13.** Numbering reptile toe clips.

The toe-clip or shell-marked numbers are tracked on a clip chart. A newly captured individual is assigned the next available number for its species and that number is marked off of the clip chart (table 1). Used numbers can be removed later from the clip chart as shown in table 1 for *Elgaria multicarinata* (ELMU), *Cnemidophorus hyperythrus* (CNHY), *Sceloporus occidentalis* (SCOC), *Sceloporus orcutti* (SCOR), and *Uta stansburiana* (UTST) , which start at higher numbers than those species which have never been captured. Alternatively, toe-clip or shell-marked numbers can be tracked on a handheld computer which removes numbers.

Toes should be clipped with very sharp scissors at the distal knuckle. To acquire more tissue, 5 to 10 mm of tail tip (of tailed specimens) may be clipped also. Applying pressure to the wound can reduce any bleeding. Submersing the animal in a jar of sand is reported to be effective also (Medica, U.S. Geological Survey, oral commun., 1995).

The toes and tail tissue can be placed into a 1.5-mL micro centrifuge tube filled with 95-percent ethanol to preserve for later genetic or skeletochronology analysis. Label the tubes with the species code, date, site, array, arm, trap, toe-clip number, and a unique identifier, such as a record identification number, if possible. Release the processed animal into nearby vegetation or cover to prevent predation and exposure. Some animals, though limbed, are unsuited for toe-clipping due to one or more aspects of the animal's anatomy. In the case of slender salamanders, their toes often are too small to distinguish one from another, let alone toe-clip.

At a new study site, captured animals will be unmarked. Record these animals as new captures and mark accordingly. On subsequent visits to the site, examine captured animals for marks. Note any marked individuals as recaptures and record the identification number of each toe-clipped or shell-marked animal.

**Figure 14.** Numbering amphibian toe clips.

**Table 1.**  Reptile and amphibian toe-clip numbers.

[Toe-clip numbers for reptiles and amphibians are listed by scientific name and species code.  For all species, as numbers are used in the field, each number is marked off, as indicated below by the strikethrough, and later removed to prevent repeating the same number. Strikethroughs indicate used toe-clip or shell-marked numbers.  Certain combinations of toe-clip numbers should be avoided, such as those including the 4 and 10 toes for amphibians and the 400 and 2,000 toes for lizards.  Certain species, such as both *Batrachoseps* species, may be incompatible with toe-clipping due to characteristics of the species and are only included to present the processing code information.  Processing codes are 1- toe-clip, weigh, measure all individuals;  2- toe-clip, weigh, measure all adults and 20 juveniles each sample period; 3- take tail tip from these.]

| Scientific Name | Species Code | Processing Code* | Toe-Clip Number | | | | | | | | | | | | | | | | |
|---|---|---|---|---|---|---|---|---|---|---|---|---|---|---|---|---|---|---|---|
| *Batrachoseps nigriventris* | BANI | 3 | | | | | | | | | | | | | | | | | |
| *Batrachoseps pacificus* | BAPA | 3 | | | | | | | | | | | | | | | | | |
| *Aneides lugubris* | ANLU | 1 | 1 | 2 | 3 | 20 | 21 | 22 | 23 | 30 | 31 | 32 | 33 | 40 | 41 | 42 | 43 | 100 | 101 |
| *Ensatina eschscholtzii* | ENES | 1 | 1 | 2 | 3 | 20 | 21 | 22 | 23 | 30 | 31 | 32 | 33 | 40 | 41 | 42 | 43 | 100 | 101 |
| *Taricha torosa* | TATO | 1 | 1 | 2 | 3 | 20 | 21 | 22 | 23 | 30 | 31 | 32 | 33 | 40 | 41 | 42 | 43 | 100 | 101 |
| *Hyla cadaverina* | HYCA | 1 | 1 | 2 | 3 | 20 | 21 | 22 | 23 | 30 | 31 | 32 | 33 | 40 | 41 | 42 | 43 | 100 | 101 |
| *Hyla regilla* | HYRE | 2 | 1 | 2 | 3 | 20 | 21 | 22 | 23 | 30 | 31 | 32 | 33 | 40 | 41 | 42 | 43 | 100 | 101 |
| *Bufo boreas* | BUBO | 2 | 1 | 2 | 3 | 20 | 21 | 22 | 23 | 30 | 31 | 32 | 33 | 40 | 41 | 42 | 43 | 100 | 101 |
| *Bufo californicus* | BUMI | 2 | 1 | 2 | 3 | 20 | 21 | 22 | 23 | 30 | 31 | 32 | 33 | 40 | 41 | 42 | 43 | 100 | 101 |
| *Bufo punctatus* | BUPU | 1 | 1 | 2 | 3 | 20 | 21 | 22 | 23 | 30 | 31 | 32 | 33 | 40 | 41 | 42 | 43 | 100 | 101 |
| *Rana aurora* | RAAU | 1 | 1 | 2 | 3 | 20 | 21 | 22 | 23 | 30 | 31 | 32 | 33 | 40 | 41 | 42 | 43 | 100 | 101 |
| *Rana catesbeiana* | RACA | 2 | 1 | 2 | 3 | 20 | 21 | 22 | 23 | 30 | 31 | 32 | 33 | 40 | 41 | 42 | 43 | 100 | 101 |
| *Scaphiopus hammondii* | SCHA | 1 | 1 | 2 | 3 | 20 | 21 | 22 | 23 | 30 | 31 | 32 | 33 | 40 | 41 | 42 | 43 | 100 | 101 |
| *Coleonyx variegatus* | COVA | 1 | 1 | 2 | 3 | 4 | 5 | 10 | 11 | 12 | 13 | 14 | 15 | 20 | 21 | 22 | 23 | 24 | 25 |
| *Xantusia henshawi* | XAHE | 1 | 1 | 2 | 3 | 4 | 5 | 10 | 11 | 12 | 13 | 14 | 15 | 20 | 21 | 22 | 23 | 24 | 25 |
| *Xantusia vigilis* | XAVI | 1 | 1 | 2 | 3 | 4 | 5 | 10 | 11 | 12 | 13 | 14 | 15 | 20 | 21 | 22 | 23 | 24 | 25 |
| *Anniella pulchra* | ANPU | 3 | 1 | 2 | 3 | 4 | 5 | 10 | 11 | 12 | 13 | 14 | 15 | 20 | 21 | 22 | 23 | 24 | 25 |
| *Elgaria multicarinata* | ELMU | 2 | 510 | 511 | 512 | 513 | 514 | 515 | 520 | 521 | 522 | 523 | 524 | 525 | 530 | 531 | 532 | 533 | 535 | 541 |
| *Eumeces gilberti* | EUGI | 1 | 1 | 2 | 3 | 4 | 5 | 10 | 11 | 12 | 13 | 14 | 15 | 20 | 21 | 22 | 23 | 24 | 25 |
| *Eumeces skiltonianus* | EUSK | 1 | 1 | 2 | 3 | 4 | 5 | 10 | 11 | 12 | 13 | 14 | 15 | 20 | 21 | 22 | 23 | 24 | 25 |
| *Cnemidophorus hyperythrus* | CNHY | 1 | 3,045 | 3,051 | 3,052 | 3,053 | 3,054 | 3,055 | 3,100 | 3,101 | 3,102 | 3,103 | 3,104 | 3,105 | 3,110 | 3,111 | 3,114 | 3,115 | 3,120 |
| *Cnemidophorus tigris* | CNTI | 1 | 1 | 2 | 3 | 4 | 5 | 10 | 11 | 12 | 13 | 14 | 15 | 20 | 21 | 22 | 23 | 24 | 25 |
| *Sceloporus occidentalis* | SCOC | 2 | 3,150 | 3,154 | 3,155 | 3,200 | 3,201 | 3,202 | 3,203 | 3,204 | 3,205 | 3,211 | 3,212 | 3,213 | 3,214 | 3,215 | 3,220 | 3,221 |
| *Sceloporus orcutti* | SCOR | 1 | 30 | 31 | 32 | 33 | 34 | 35 | 40 | 41 | 42 | 43 | 44 | 45 | 50 | 51 | 52 | 53 | 54 |
| *Uta stansburiana* | UTST | 2 | 3,202 | 3,212 | 3,214 | 3,215 | 3,220 | 3,224 | 3,225 | 3,230 | 3,231 | 3,232 | 3,233 | 3,234 | 3,235 | 3,240 | 3,241 | 3,242 | 3,243 |
| *Phrynosoma coronatum* | PHCO | 1 | 1 | 2 | 3 | 4 | 5 | 10 | 11 | 12 | 13 | 14 | 15 | 20 | 21 | 22 | 23 | 24 | 25 |
| *Gambelia wislizenii* | GAWI | 1 | 1 | 2 | 3 | 4 | 5 | 10 | 11 | 12 | 13 | 14 | 15 | 20 | 21 | 22 | 23 | 24 | 25 |

## 6.5.2 Processing Snakes and Other Limbless Specimens

To process limbless reptiles, first record the array and trap in which the animal is captured. Record the species, age (juvenile/adult), any unusual markings, deformities, and (or) injuries. Record the sex and reproductive status, if possible. Measure the length by placing a ruler on the ventral side of the animal and measuring the length from the tip of the snout to the vent. A measuring tape may be needed when measuring most adult snakes. Length is measured to the nearest millimeter. Before weighing an animal, make sure that the spring scale has been tared to correct for the weight of the weighing bag. Record the weight as accurately as the spring scale allows.

Most snakes are marked by scale-clipping. Venomous snakes (see below), blind snakes, and legless lizards typically are not marked. Snakes are marked by clipping their post-anal scales such that no two individuals of a species at a study site are marked with the same number (fig. 15). The post-anal scales are located on the ventral side of a snake's tail, posterior to the vent or anal plate. Some snakes have divided post-anal scales while some have undivided post-anal scales. The scales are assigned numbers by looking at the individual from the bottom or ventral side with the head up in the 12 o'clock position. For snakes with divided post-anal scales, the two columns of scales are numbered 1 to 9 beginning with the scales closest to the vent and working towards the end of the tail. The left side represents 100s and the right side represents 10s. Snakes with undivided post-anal scales are treated like snakes with divided scales. The left and right halves of the same scale are clipped in the same manner as separate left and right scales. The first two scales on each side (10, 20, 100, and 200) are not clipped to reduce the chance of infection near the vent area.

All species (except venomous snakes) are inspected for markings upon capture. Marked individuals are noted as recaptures. A newly captured individual is assigned a new number in sequence. The number is marked as used on the scale-clip chart and later removed (table 2). Alternatively, scale-clip numbers can be tracked on a handheld computer with numbers removed automatically after use.

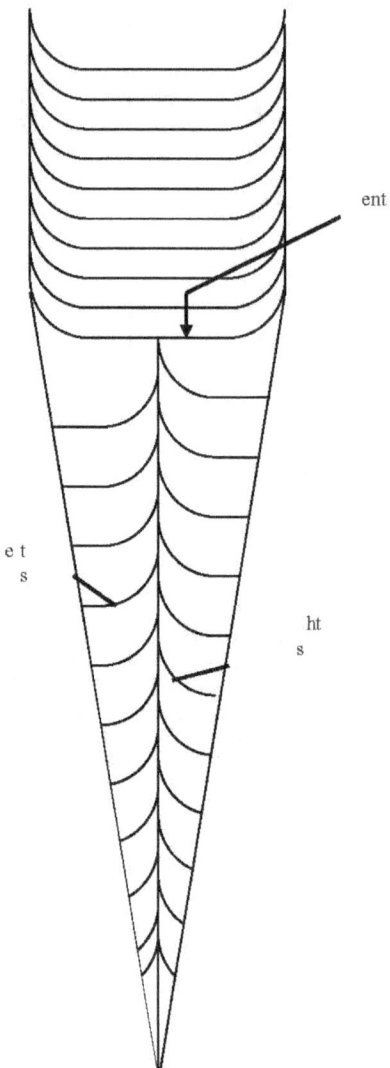

**Figure 15.**    Numbering snake scale clips.

**Table 2.** Snake scale-clip numbers.

[Scale-clip numbers for snakes are listed by scientific name and species code. For all species, as numbers are used in the field, the number is marked off and removed to prevent repeating the same number. Certain combinations of scale-clip numbers should be avoided, such as combinations including the 1 and 2 scale. Certain species, such as Leptotyphlops humilis, may be incompatible with scale-clipping due to characteristics of the species and are only included to present the processing code information. 1- Not individually tail-clipped, but scales along the side of the tail removed for scaring and saved for tissue. 2- Scales under tail not divided, but remove either from right or left sides. 3- Scales divided, and easy to clip. Also remove 0.5- to 1-inch of tail tip for tissue. 4- Scales divided, but hard to clip. Also remove 0.5- to 1-inch of tail tip for tissue. 5- Do not clip scales, weigh, or measure, release, and record species and approximate size. Snake scale-clip numbers are recorded in the data book with the left side as the 100s digit and the right side the 10s (that is, 3L 4R would be recorded as 340 for scale-clip number).]

| Scientific name | Species code | Processing code | L | R | L | R | L | R | L | R | L | R | L | R | L | R | L | R | L | R | L | R | L | R | L | R | L | R |
|---|---|---|---|---|---|---|---|---|---|---|---|---|---|---|---|---|---|---|---|---|---|---|---|---|---|---|---|---|
| | | | | | | | | | | | | | Scale-clip number | | | | | | | | | | | | | | |
| Leptotyphlops humilis | LEHU | 1 | 3 | 3 | 3 | 4 | 3 | 5 | 3 | 6 | 3 | 7 | 3 | 8 | 3 | 9 | 4 | 4 | 4 | 5 | 4 | 6 | 4 | 7 | 4 | 8 | 4 | 9 |
| Lichanura trivirgata | LITR | 2 | 3 | 3 | 3 | 4 | 3 | 5 | 3 | 6 | 3 | 7 | 3 | 8 | 3 | 9 | 4 | 4 | 4 | 5 | 4 | 6 | 4 | 7 | 4 | 8 | 4 | 9 |
| Arizona elegans | AREL | 3 | 3 | 3 | 3 | 4 | 3 | 5 | 3 | 6 | 3 | 7 | 3 | 8 | 3 | 9 | 4 | 4 | 4 | 5 | 4 | 6 | 4 | 7 | 4 | 8 | 4 | 9 |
| Charina bottae | CHBO | 2 | 3 | 3 | 3 | 4 | 3 | 5 | 3 | 6 | 3 | 7 | 3 | 8 | 3 | 9 | 4 | 4 | 4 | 5 | 4 | 6 | 4 | 7 | 4 | 8 | 4 | 9 |
| Coluber constrictor | COCO | 3 | 3 | 3 | 3 | 4 | 3 | 5 | 3 | 6 | 3 | 7 | 3 | 8 | 3 | 9 | 4 | 4 | 4 | 5 | 4 | 6 | 4 | 7 | 4 | 8 | 4 | 9 |
| Diadophis punctatus | DIPU | 4 | 3 | 3 | 3 | 4 | 3 | 5 | 3 | 6 | 3 | 7 | 3 | 8 | 3 | 9 | 4 | 4 | 4 | 5 | 4 | 6 | 4 | 7 | 4 | 8 | 4 | 9 |
| Hypsiglena torquata | HYTO | 3 | 3 | 3 | 3 | 4 | 3 | 5 | 3 | 6 | 3 | 7 | 3 | 8 | 3 | 9 | 4 | 4 | 4 | 5 | 4 | 6 | 4 | 7 | 4 | 8 | 4 | 9 |
| Lampropeltis getula | LAGE | 3 | 3 | 3 | 3 | 4 | 3 | 5 | 3 | 6 | 3 | 7 | 3 | 8 | 3 | 9 | 4 | 4 | 4 | 5 | 4 | 6 | 4 | 7 | 4 | 8 | 4 | 9 |
| Lampropeltis zonata | LAZO | 3 | 3 | 3 | 3 | 4 | 3 | 5 | 3 | 6 | 3 | 7 | 3 | 8 | 3 | 9 | 4 | 4 | 4 | 5 | 4 | 6 | 4 | 7 | 4 | 8 | 4 | 9 |
| Masticophis flagellum | MAFL | 3 | 3 | 3 | 3 | 4 | 3 | 5 | 3 | 6 | 3 | 7 | 3 | 8 | 3 | 9 | 4 | 4 | 4 | 5 | 4 | 6 | 4 | 7 | 4 | 8 | 4 | 9 |
| Masticophis lateralis | MALA | 3 | 3 | 3 | 3 | 4 | 3 | 5 | 3 | 6 | 3 | 7 | 3 | 8 | 3 | 9 | 4 | 4 | 4 | 5 | 4 | 6 | 4 | 7 | 4 | 8 | 4 | 9 |
| Pituophis catenifer | PIME | 3 | 3 | 3 | 3 | 4 | 3 | 5 | 3 | 6 | 3 | 7 | 3 | 8 | 3 | 9 | 4 | 4 | 4 | 5 | 4 | 6 | 4 | 7 | 4 | 8 | 4 | 9 |
| Rhinocheilus lecontei | RHLE | 3 | 3 | 3 | 3 | 4 | 3 | 5 | 3 | 6 | 3 | 7 | 3 | 8 | 3 | 9 | 4 | 4 | 4 | 5 | 4 | 6 | 4 | 7 | 4 | 8 | 4 | 9 |
| Salvadora hexalepis | SAHE | 3 | 3 | 3 | 3 | 4 | 3 | 5 | 3 | 6 | 3 | 7 | 3 | 8 | 3 | 9 | 4 | 4 | 4 | 5 | 4 | 6 | 4 | 7 | 4 | 8 | 4 | 9 |
| Tantilla planiceps | TAPL | 4 | 3 | 3 | 3 | 4 | 3 | 5 | 3 | 6 | 3 | 7 | 3 | 8 | 3 | 9 | 4 | 4 | 4 | 5 | 4 | 6 | 4 | 7 | 4 | 8 | 4 | 9 |
| Thamnophis elegans | THEL | 3 | 3 | 3 | 3 | 4 | 3 | 5 | 3 | 6 | 3 | 7 | 3 | 8 | 3 | 9 | 4 | 4 | 4 | 5 | 4 | 6 | 4 | 7 | 4 | 8 | 4 | 9 |
| Thamnophis hammondii | THHA | 3 | 3 | 3 | 3 | 4 | 3 | 5 | 3 | 6 | 3 | 7 | 3 | 8 | 3 | 9 | 4 | 4 | 4 | 5 | 4 | 6 | 4 | 7 | 4 | 8 | 4 | 9 |
| Thamnophis sirtalis | THSI | 3 | 3 | 3 | 3 | 4 | 3 | 5 | 3 | 6 | 3 | 7 | 3 | 8 | 3 | 9 | 4 | 4 | 4 | 5 | 4 | 6 | 4 | 7 | 4 | 8 | 4 | 9 |
| Trimorphodon biscutatus | TRBI | 3 | 3 | 3 | 3 | 4 | 3 | 5 | 3 | 6 | 3 | 7 | 3 | 8 | 3 | 9 | 4 | 4 | 4 | 5 | 4 | 6 | 4 | 7 | 4 | 8 | 4 | 9 |
| Crotalus mitchellii | CRMI | 5 | 3 | 3 | 3 | 4 | 3 | 5 | 3 | 6 | 3 | 7 | 3 | 8 | 3 | 9 | 4 | 4 | 4 | 5 | 4 | 6 | 4 | 7 | 4 | 8 | 4 | 9 |
| Crotalus ruber | CRRU | 5 | 3 | 3 | 3 | 4 | 3 | 5 | 3 | 6 | 3 | 7 | 3 | 8 | 3 | 9 | 4 | 4 | 4 | 5 | 4 | 6 | 4 | 7 | 4 | 8 | 4 | 9 |
| Crotalus viridis | CRVI | 5 | 3 | 3 | 3 | 4 | 3 | 5 | 3 | 6 | 3 | 7 | 3 | 8 | 3 | 9 | 4 | 4 | 4 | 5 | 4 | 6 | 4 | 7 | 4 | 8 | 4 | 9 |

To perform a scale clip, use small sharp scissors to clip the outer corners of the scales. It is important to clip deeply enough so that the scute does not regenerate. Extra scales on the belly or 5 to 10 mm of tail tip can be clipped to acquire extra tissue from medium and large snakes for genetic material. Five to 10 mm of tail tip can be collected from animals with scales too small to clip, such as small snakes and limbless lizards. Place the clipped scales and (or) tail tip into a 1.50-mL microcentrifuge tube filled with 95 percent ethanol. Label the tubes with the species, date, site, array, arm, trap, scale-clip number, and a unique identifier, such as a record ID #, if possible. Finally, the animal can be released into nearby vegetation or cover.

Due to safety concerns with large crews of field technicians, venomous snakes are not weighed or measured. Length, age, and sex may be approximated without handling the animals. Venomous snakes in pitfall traps can be removed carefully by using a snake stick if they are not large enough to get out themselves. Venomous snakes can be removed from funnel traps by carefully removing one end of the trap, tipping the snake out a few meters away from the array, and allowing it to find cover away from the researcher.

## 6.5.3 Incidental Captures and Observations

Many non-targeted species may be trapped or observed on site incidentally. How these animals are recorded and processed depends on the research objectives. For our studies, the focal animal types are herpetofauna. Small mammals and invertebrates captured in the pitfall traps are treated as incidentals. We identify small mammals to the species level, if possible, and record the observation. Relative age and gender of specimens are noted also. Live animals are released and dead animals are collected as vouchers. Direct observations of megafauna at study sites are recorded daily while traps are being sampled. The term "megafauna" is used to refer to any animal other than the reptiles, amphibians, and small mammals normally found in the pitfall traps. For our purposes, this includes ground birds, large carnivores, and large herbivores observed at the site. Direct observations and signs from megafauna, such as tracks and scat, may be recorded also. Invertebrates are not identified in the field but

are collected from pitfall traps on the last day of sampling. Forceps are used to collect the animals and put them into an appropriate container filled with 70-percent ethanol for storage and future analysis.

## 6.6 Vegetation Surveys

Vegetation is surveyed at each pitfall trap array through the use of a point-intercept transect technique (Sawyer and Keeler-Wolf, 1995) and recorded on a vegetation data sheet (table 3). Data are collected for species composition, canopy height, substrate, leaf litter depth, as well as incidental plant species, slope and aspect.

A 50-m transect is sampled for each array and data are recorded at points every 0.5 m along the transect. To perform a survey transect, run a 50-m line that is centered on the midpoint of the array (center bucket) in a north-south orientation. The line will extend 25 m north and 25 m south of the center bucket. Begin at the northernmost point of the transect. Line 1 of the data sheet always refers to the northern-most point of the transect. Place a telescoping measuring rod vertical to the ground at each 0.5-m survey point. First, record the tallest vegetation height (canopy height) on the vegetation data sheet (table 3). Record the plant species under one of three height classifications: tree, shrub, or herb. Tree refers to any plant taller then 3 m. Shrub refers to any plant between 0.5- and 3-m. Herb refers to any plant less than 0.5 cm. After recording the tallest plant height and species, record any additional species that touch the telescoping rod and their height classifications (without recording numerical height). At each point, also record substrate type and leaf litter depth. After completing the transect, record any incidental plant species around the array that were not recorded at any point on the transect. For each array location, note the slope, aspect, GPS coordinates, and date.

These transects are completed once at each array unless there is a substantial change in vegetation, usually due to disturbance such as fire. In such an instance, the transect technique may be repeated. Vegetation surveys may be performed more frequently depending on the nature of the research and the questions they are intended to answer.

**Table 3.**   Vegetation data spreadsheet.

[At each pitfall array, vegetation data are collected at 0.5-meter intervals along a 50-meter north/south transect, centered on the center pitfall trap. In cases where the data will be available to the public, asterisks may be used to show the precision of the coordinate positions were reduced to protect the equipment in the field. The four letter species codes consist of the first two letters of the genus and the species.  See appendix 2 for a translation of species codes and descriptions of substrate abbreviations. location date, date when the coordinate data was collected; cm, centimeter; m, meter]

| Site Name: POINT LOMA | | | Array #: 13 | | Location:  32.669** N 117.241*** W | | | | | |
|---|---|---|---|---|---|---|---|---|---|---|
| Slope: (N-S) -30, -25 | | (E-W) -7, -3 | | | Location date:  01-24-96 | | | Transect date  01-24-96 | | |
| Incidental Species | | | | | | | | | | |

| N | Canopy height (m) | Species | | | | | | | | Litter depth (cm) |
|---|---|---|---|---|---|---|---|---|---|---|
| | | Tree | Tree | Shrub | Shrub | Shrub | Herb | Herb | Substrate | |
| 1 | 0.19 | | | | | | DULA | ARCA | OR | 0.0 |
| 2 | 0.00 | | | | | | | | SS | 0.0 |
| 3 | 0.00 | | | | | | | | SS | 0.0 |
| 4 | 0.00 | | | | | | | | SS | 0.0 |
| 5 | 0.00 | | | | | | | | BR | 0.0 |
| 6 | 0.44 | | | | | | ARCA | | LL | 0.5 |
| 7 | 0.49 | | | | | | ARCA | | LL | 0.5 |
| 8 | 0.47 | | | | | | ARCA | ENCA | LL | 0.5 |
| 9 | 0.50 | | | ARCA | ENCA | | | | LL | 0.5 |
| 10 | 0.53 | | | ARCA | ENCA | | NNG | | OR | 0.0 |
| 11 | 0.56 | | | ARCA | | | | | LL | 0.5 |
| 12 | 0.60 | | | ARCA | | | | | LL | 1.0 |
| 13 | 0.73 | | | ARCA | | | | | LL | 1.0 |
| 14 | 0.70 | | | ARCA | | | | | LL | 1.0 |
| 15 | 0.60 | | | ARCA | | | | | LL | 2.0 |
| 16 | 0.00 | | | | | | | | LL | 11.0 |
| 17 | 0.00 | | | | | | | | LL | 8.0 |
| 18 | 0.20 | | | | | | ARCA | | LL | 4.0 |
| 19 | 0.31 | | | | | | ARCA | | LL | 4.0 |
| 20 | 0.74 | | | RHIN | | | | | SS | 0.0 |
| 21 | 0.90 | | | RHIN | | | | | LL | 0.5 |
| 22 | 1.05 | | | RHIN | ARCA | | | | LL | 0.5 |
| 23 | 0.71 | | | ARCA | | | | | LL | 1.0 |
| 24 | 0.72 | | | ARCA | | | | | LL | 1.0 |
| 25 | 0.83 | | | ARCA | | | | | LL | 3.0 |
| 26 | 0.63 | | | ARCA | | | | | LL | 3.0 |
| 27 | 0.97 | | | RHIN | ARCA | | | | LL | 1.0 |
| 28 | 0.60 | | | RHIN | | | | | LL | 2.0 |
| 29 | 0.95 | | | RHIN | | | | | LL | 2.0 |
| 30 | 0.81 | | | RHIN | | | | | LL | 2.0 |
| 31 | 0.53 | | | RHIN | | | | | LL | 2.0 |
| 32 | 0.75 | | | RHIN | | | | | LL | 3.0 |
| 33 | 0.75 | | | ERFA | RHIN | | | | LL | 3.0 |
| 34 | 0.70 | | | RHIN | ERFA | | | | LL | 3.0 |
| 35 | 0.84 | | | RHIN | | | | | LL | 2.0 |

# 7.0 Survey Data Management

## 7.1 Data Collection and Entry

Data are recorded onto standardized data forms on either paper or on a handheld computer. Data collected in the field can be grouped into one of several categories, such as animals captured in the pitfall arrays, megafauna, or weather. When collecting data on paper, each category has a separate form. On the handheld computer, the "megafauna" category has been incorporated into the animal data form.

### 7.1.1 Pitfall Capture Data

The majority of data fields recorded for each captured animal are the same whether data are collected on paper or on the handheld computer (figs. 16 and 17).

Data fields and definitions are

Date: recorded as the present date. On the handheld computer, this field automatically is generated based on the handheld's preset, internal clock.

Site Name: referring to the overall study area and its components. Those collecting data on the handheld computers need to enter the site name only at the start of the fieldwork; it will automatically be carried over to subsequent records until the collector changes the site name.

Array Number: within a site, there are multiple pitfall/funnel trap study plots, each with an individual identification number.

Bucket Number: each pitfall trap (fig. 1) has a preassigned, unique identification based on the arm along which it occurs and its position along that arm. Using the handheld computer, the collector selects the bucket number from a predefined list of entries.

Snake Trap #: as with Bucket Number, each funnel/snake trap has a preassigned, unique number based on the arm along which it is located. To facilitate entry, a pop-up list (fig. 17) is used on the handheld to enter the snake trap number.

Species: entered as a four-letter code, consisting of the first two letters of both the genus and the species. Examples of these codes would be SCOC for the western fence lizard (*Sceloporus occidentalis*) and RHLE for the long-nosed snake (*Rhinocheilus lecontei*). To aide in data entry on the handheld computers, species names and codes have been predefined and appear as a pop-up list from which the collector selects the desired species code. The entry of a species code is further aided by an additional data field on the handheld computer described below.

Sex: M / F / ?, (Male, Female, Unknown) as determined by physical characteristics, appearing as a pop-up list on the handheld computer.

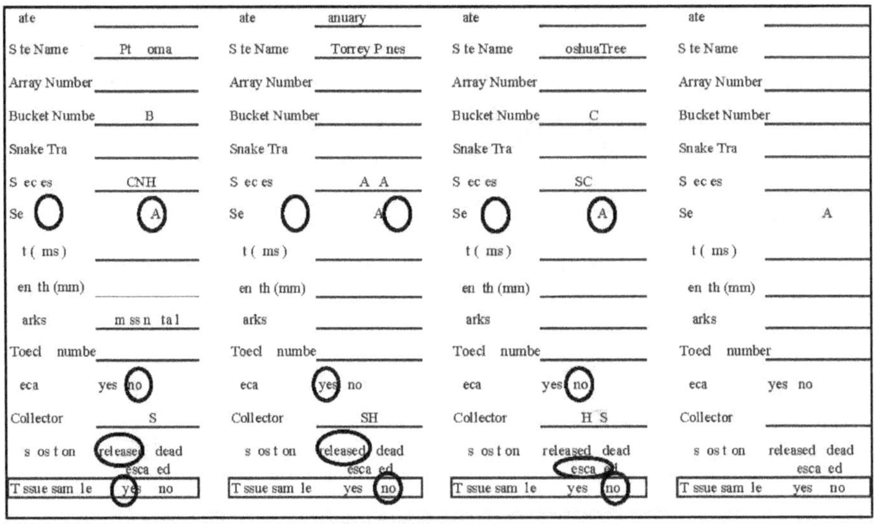

| | | | | | | | |
|---|---|---|---|---|---|---|---|
| ate | | ate | anuary | ate | | ate | |
| S te Name | Pt oma | S te Name | Torrey P nes | S te Name | oshuaTree | S te Name | |
| Array Number | | Array Number | | Array Number | | Array Number | |
| Bucket Numbe | B | Bucket Number | | Bucket Numbe | C | Bucket Number | |
| Snake Tra | | Snake Tra | | Snake Tra | | Snake Tra | |
| S ec es | CNH | S ec es | A A | S ec es | SC | S ec es | |
| Se | A | Se | A | Se | A | Se | A |
| t ( ms) | | t ( ms) | | t ( ms) | | t ( ms) | |
| en th (mm) | | en th (mm) | | en th (mm) | | en th (mm) | |
| arks | m ss n ta l | arks | | arks | | arks | |
| Toecl numbe | | Toecl numbe | | Toecl numbe | | Toecl number | |
| eca | yes no | eca | yes no | eca | yes no | eca | yes no |
| Collector | S | Collector | SH | Collector | H S | Collector | |
| s os t on | released dead esca ed | s os t on | released dead esca ed | s os t on | released dead esca ed | s os t on | released dead esca ed |
| T ssue sam le | yes no | T ssue sam le | yes no | T ssue sam le | yes no | T ssue sam le | yes no |

**Figure 16.** Paper data form for recording animal captures.

Age: A / J / ?, (Adult, Juvenile, Unknown) which can be established by size and appearance, appearing as a pop-up list on the handheld computer.

Weight: the weight is recorded in grams, as accurately as the spring scale displays.

Length: as measured from snout to vent for lizards, snakes, and salamanders, or snout to urostyle for frogs and toads. Record carapace length for turtles or tortoises. Record all length measurements in millimeters.

Marks or Notes: any unusual markings, injuries, deformities, and reproductive status can be recorded in this data field. This field also allows for the documentation of any outstanding features of the specimen. For example, the amount of tail regeneration in lizards, scars, or additional missing toes may be recorded for general information. For handheld computer users, a list of commonly used phrases can be accessed to minimize manual input on the computer.

Toe-clip number: listed as a four-digit figure, for consistency, even if only one toe on one foot has been cut. Number 5 would be documented on the paper data sheet or handheld computer as 0005. If study sites are sufficiently far apart, toe-clip, scale-clip, and shell-marked numbers are not carried over from one site to the next. Each site has its own independent set of numbers.

Recap: Y / N / ?, (Yes, No, Unknown) the question mark often is used in the case of amphibians. Recaptures of some amphibians can be identified accurately within a sample period. However, since some species can regenerate toes, it may be impossible to tell whether they have been trapped in previous sample periods, based on these criteria. "Unknown" may indicate animals that have lost toes naturally. Each recap needs to be evaluated carefully.

Collector: typically recorded as the initials of the observer(s). An effort should be made to ensure that each observer has unique initials, using the middle initial as necessary. Full names with a key for initials should be maintained in the laboratory.

Disposition: released / dead / escaped. Once all relevant data have been gathered, most animals, if alive, are freed. If an animal is found dead in the traps, weight and length are still recorded and the body is saved for a voucher and tissue sample. If the specimen is in poor condition, these measurements can be skipped. Occasionally, an animal is resourceful enough to escape from the hand of the researcher, resulting in incomplete data for that animal.

Tissue: yes / no. The processing codes, as explained in tables 1 and 2, outline whether or not a tissue sample is collected from an animal. In addition to the toes and scales that are clipped to tag the specimens, a piece of the tail may be collected for the purpose of a tissue sample. For turtles, the tail tip will be the only tissue sample collected. A new tissue sample will not be collected if an animal is a recapture.

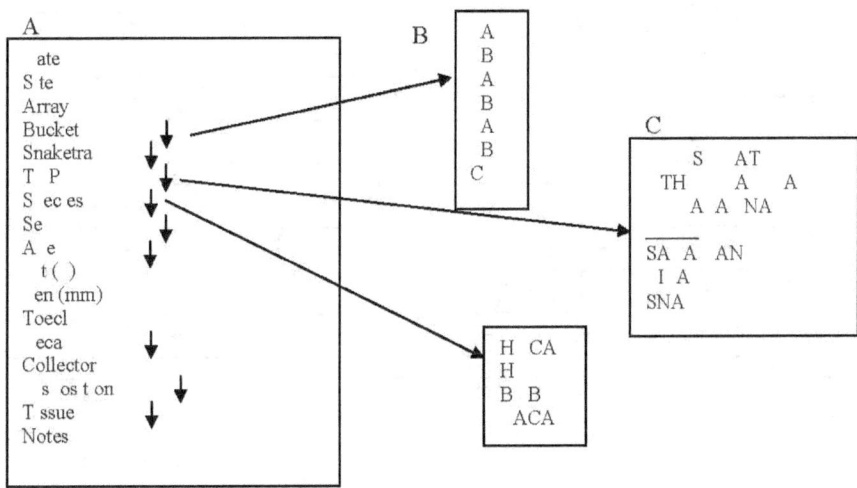

**Figure 17.**    Handheld computer form for recording animal capture data.

In addition to the data fields described above, handheld computer users have one more field to enter, "TYPE". On the handheld computer, just above "SPECIES", is "TYPE", which is a pop-up list used to identify the kind of animal that is being processed (fig. 17C). The use of the "TYPE" data field acts to reduce the number of species codes from which the handheld user has to choose when entering the species code in the "SPECIES" field. For example, when the user chooses "FROG" from the "TYPE" pop-up list, only a list of frog species will appear in the "SPECIES" pop-up list. Likewise, when "SNAKE" is selected under "TYPE", only snake species will be shown in the pop-up list for "SPECIES". There are type and species codes for lizards, salamanders, and small mammals. The list of options under "TYPE" can be modified to include other categories of animals, as necessary for the study.

## 7.1.2 Megafauna Data

Megafauna observations are recorded onto standardized megafauna forms on a daily basis on either paper or a handheld computer (fig. 18). The term "Megafauna" is used here to refer to any animal other than the reptiles, amphibians, and small mammals that are found in the pitfall traps. Signs from megafauna, such as tracks and scat, also may be recorded. Handheld computer users can record the megafauna records in a standard animal form by making a check mark next to the appropriate species.

## 7.1.3 Weather Data

Weather observations are recorded onto standardized weather data forms on a daily basis on either paper data forms or on a handheld computer (see figs. 19 and 20). Air temperatures at a site may be recorded using a "Max/Min" thermometer or a temperature sensitive data logger. When collecting data on paper, the name of the site is recorded across the top of the page. The date and related weather variables are recorded in columns. When using the "Max/Min" thermometers, the high and low temperatures over the previous 24 hours are recorded, along with the array number and site. The "Max/Min" thermometers must be reset daily. When using the data loggers, the unit is put into position on the opening day of the sample period and left alone for the duration. In addition to temperatures, a general weather condition is recorded daily: sunny, foggy, overcast, precipitation. The data loggers used to record environmental temperatures must be set up and downloaded using a computer and the related application (for example, HOBO data logger with Boxcar application). In our protocol, the loggers are programmed to gather temperature data every 15 minutes. At the end of the sample period, the logger is returned to the lab to be downloaded. Once the data have been downloaded, a graph of the temperatures during the sample period can be generated (fig. 21).

The weather data form on the handheld computer is formatted in conjunction with the temperature data loggers. Therefore, the data fields are slightly different from those on the paper weather-data form. On the handheld computer, the

| S te | Pt oma | | S te | | | S te | | |
|---|---|---|---|---|---|---|---|---|
| Sam le Per od Start | ate | | Sam le Per od Start | ate | | Sam le Per od Start | ate | |
| bserver | S | | bserver | | | bserver | | |
| | Present | | | Present | | | Present | |
| ule deer | | | ule deer | | | ule deer | | |
| Coyote | | | Coyote | | | Coyote | | |
| rey o | | | rey o | | | rey o | | |
| ed o | | | ed o | | | ed o | | |
| Bobcat | | | Bobcat | | | Bobcat | | |
| ounta n on | | | ounta n on | | | ounta n on | | |
| Bad er | | | Bad er | | | Bad er | | |
| ackrabb t | | | ackrabb t | | | ackrabb t | | |
| rey s u rrel | | | rey s u rrel | | | rey s u rrel | | |
| round s u rrel | | | round s u rrel | | | round s u rrel | | |
| oad runner | | | oad runner | | | oad runner | | |
| ua l | array | | ua l | | | ua l | | |
| Burrow n wl | | | Burrow n wl | | | Burrow n wl | | |
| ther round B rd | | | ther round B rd | | | ther round B rd | | |
| ther ammals | | | ther ammals | | | ther ammals | | |

**Figure 18.** Paper data form for recording megafauna captures.

data fields are Site, Date, Conditions Start, Conditions End, and Notes (fig. 20). On the day that the site is opened, the user must enter the name of the site. This name automatically will be carried over to the following weather records until the user changes the name of the site. The "Date" field is generated automatically based on the hand-held computer's internal clock. "Conditions Start" and "Conditions End" are predefined pop-up lists from which the user can select a weather condition based on the same four categories described above: Sunny, foggy, overcast, precipitation. In the "Notes" field, those using handheld computers can record which data logger was used in the field and its corresponding array location.

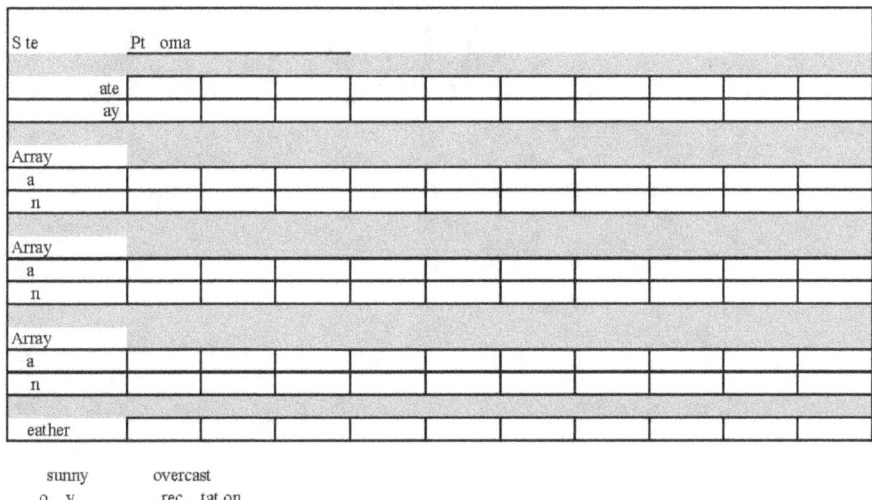

**Figure 19.**   Paper data form for recording weather conditions.

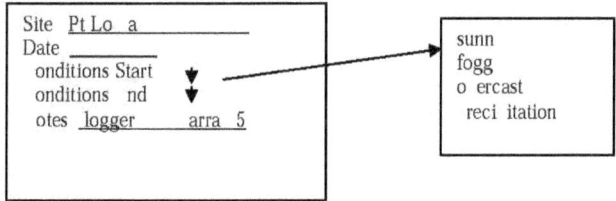

**Figure 20.**   Handheld computer data form for recording weather conditions.

ate          Tem erature
                e  rees Cels us

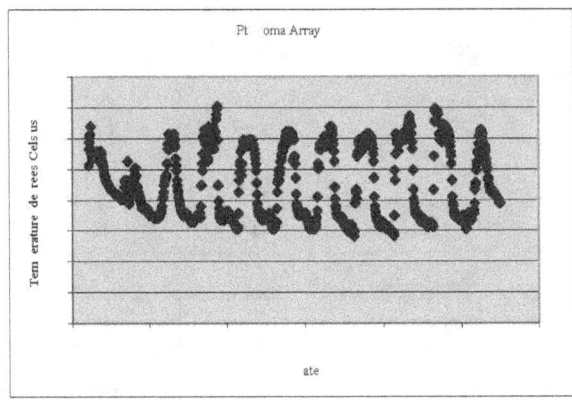

**Figure 21.**   Tabular and graphical displays of temperature patterns for a sampling period.

## 7.2 Quality Assurance/Quality Control

Multiple levels of review ensure a high standard of quality assurance and control of field data. First is a same-day review of the data sheets (or handheld forms) by the field technician. This should be done immediately after returning from the field while the captures from the field day are still fresh in the technician's mind. A second review of the data is conducted after the sample period when entering the data into a spreadsheet to go into the main database. Data are proofed to ensure that the spreadsheet data accurately mirrors the data collected from the field.

Additional proofing of the data involves checking the accuracy of the toe-clip numbers. The species records and toe-clip charts are compared to ensure that all numbers that were used in the field were marked as used and removed from the list. Next, toe-clip numbers are checked for replication. If the same toe-clip number was used twice for the same species, this is documented in the notes field of each animal. Lastly, the data are checked to make sure that the correct toe-clip numbers were used with the correct species. Any abnormalities are checked with the field technician, noted, and (or) corrected.

The animal data are proofed also, which includes reviewing species identifications and measurement data. Questions such as "Are weight and length measurements appropriate for the species?" or "Was a decimal point put in the wrong place?" are answered. The tissue vials are checked for accuracy against the animal records.

The remainder of the data are then checked for completeness and accuracy. For example, the data are checked to verify that each sample day of the sample period is represented in the data. If a date is not accounted for, the data are reviewed to determine if there were no captures for that day. If, indeed, there were no animal captures for a day in the sample period, a new record is added to the data set that consists of the date, the name of the site, the field technician, and the note "No Captures". This record is needed to maintain an accurate account of the effort that has been put forth when calculating capture rates.

Once the data have been checked for accuracy, they are sorted by animal type (herpetofauna, small mammals, and megafauna). The data for each category of animal are added to the main database for each animal type.

## 7.3 Data Organization and Summarization

The data organization and summarization methods presented here are based on the use of Microsoft Excel spreadsheets and the tools and functions available within them. Newer data storage programs are available and should be reviewed for their potential advantages.

The reptile and amphibian data for the site are stored as Microsoft Excel files. Examples of raw data files for animal records are given in tables 4 and 5. A map of the study site showing the general location of the arrays can be generated using a topographic mapping program such as TOPO! (Wildflower Productions, San Francisco, California) or other geographic information system program (fig. 22). Other metadata important to maintain includes the latitude and longitude of each array, the sampling history of the site, the number of days the site has been sampled, and a brief description of the site (table 6).

Tables in Microsoft Excel are useful for reviewing and summarizing survey data. "Species by Array Tables" can be generated to count the number of times that each species has been documented at each array. We generate "Species by Array Tables" for the small mammals and herpetofauna at each study site (tables 7 and 8, respectively). Tables can be generated to analyze the captures for a single sampling period or from multiple sampling periods. They can be manipulated to reflect specific conditions in any of the data fields, such as removing and calculating the number of recaptures.

The reptile and amphibian species detected using our pitfall trapping technique can be compared with the species known to have existed previously at a study site. For example in Point Loma, near San Diego, California, a significant record exists for just such a comparison (L.M. Klauber, San Diego Natural History Museum, unpub. field notes, 1923–40). After several years of surveying, we have not captured five of the snake species and two of the lizard species that were known to occur at the site historically. These seven species of reptiles represent over 30 percent of the species documented at Point Loma in the 1940s. While it is difficult to say for sure that these species have been extirpated from the study site, it can be argued that they no longer are present. First, to date, there have been 320 sample days at this study site (table 6) containing 17 pitfall trap arrays. With seven pitfall traps and three funnel traps per array, this is equivalent to 38,080 pitfall trap days and 16,320 funnel trap days. This represents a significant sampling effort using a technique that has successfully captured these species at other study sites. Second, targeted searches were conducted using specialized techniques that still did not detect these species. These types of historical comparisons are important for documenting local and regional species declines. They serve to bring attention to species that may be particularly vulnerable to human disturbance or habitat fragmentation. These analyses may stimulate further research, monitoring, and protection for focal species and aid those responsible for making land- and resource-management decisions.

Additional summarization may include calculating an index of relative abundance for each species between arrays within a study site or among study sites. The comparison within a site can be accomplished by dividing the data in table 8 (with animals recaptured within each sample period removed) by the number of sampling days from table 6.

**Table 4.**   Reptile and amphibian data spreadsheet.

[Reptile and amphibian data as transcribed from a paper data sheet are stored in a Microsoft Excel spreadsheet. See the appropriate section within appendix 2 for the abbreviations or codes and their explanations. g, gram; mm, millimeter; CNHY, *Cnemidophorus hyperythrus*; SCOC, *Sceloporus occidentalis*; ELMU, *Elgaria multicarinata*; CRVI, *Crotalus viridis*; UTST, *Uta stansburiana*; R, released; D, dead; M, male; F, female; ?, unknown; A, adult; J, juvenile; N, no; Y, yes; SB, Shane Bagnall; SW, Samantha Weber]

| Date | Site name | Array No. | Bucket | Snake trap | Species | Sex | Age | Weight (g) | Length (mm) | Toe-clip No. | Recap (?) | Collector's initials | Dispo-sition | Tissue (?) | Sample No. | Notes |
|---|---|---|---|---|---|---|---|---|---|---|---|---|---|---|---|---|
| 08-01-95 | Point Loma | 1 | 2B | | CNHY | M | A | 6.0 | 58 | 0001 | N | SB-SW | R | Y | 1 | |
| 08-01-95 | Point Loma | 2 | 3B | | SCOC | ? | J | 0.6 | 26 | 0001 | N | SB-SW | R | Y | 1 | |
| 08-01-95 | Point Loma | 4 | 1B | | ELMU | ? | A | 24.0 | 122 | 0001 | N | SB-SW | R | Y | 1 | |
| 08-01-95 | Point Loma | 5 | | 3 | CRVI | ? | A | | | | N | SB-SW | R | N | 1 | Trapped with live shrew |
| 08-01-95 | Point Loma | 5 | 3B | | SCOC | F | A | 10.0 | 65 | 0004 | N | SB-SW | R | N | 1 | |
| 08-01-95 | Point Loma | 5 | 3B | | SCOC | M | A | 12.0 | 67 | 0005 | N | SB-SW | R | Y | 1 | |
| 08-01-95 | Point Loma | 6 | 2A | | SCOC | ? | J | 0.8 | 25 | 0010 | N | SB-SW | R | Y | 1 | |
| 08-01-95 | Point Loma | 6 | 3A | | UTST | M | A | 3.5 | 46 | 0001 | N | SB-SW | R | Y | 1 | |
| 08-01-95 | Point Loma | 6 | 3B | | UTST | M | A | 4.5 | 50 | 0002 | N | SB-SW | R | Y | 1 | |
| 08-01-95 | Point Loma | 8 | 1B | | ELMU | ? | A | 34.0 | 114 | 0002 | N | SB-SW | R | Y | 1 | |
| 08-01-95 | Point Loma | 9 | 1 | | CNHY | F | A | 6.0 | 55 | 0002 | N | SB-SW | R | Y | 1 | |
| 08-01-95 | Point Loma | 10 | C | | SCOC | ? | J | 0.8 | 23 | 0011 | N | SB-SW | R | Y | 1 | |
| 08-01-95 | Point Loma | 11 | 1B | | SCOC | ? | J | 1.0 | 27 | 0012 | N | SB-SW | R | Y | 1 | |
| 08-01-95 | Point Loma | 13 | 1B | | SCOC | ? | J | 0.5 | 29 | 0002 | N | SB-SW | R | Y | 1 | |
| 08-01-95 | Point Loma | 13 | C | | SCOC | M | A | 8.0 | 63 | 0003 | N | SB-SW | R | Y | 1 | |
| 08-02-95 | Point Loma | 3 | 3C | | CNHY | F | A | 4.5 | 61 | 0003 | N | SB-SW | R | Y | 1 | |
| 08-02-95 | Point Loma | 4 | 1A | | ELMU | ? | J | 14.0 | 87 | 0003 | N | SB-SW | R | Y | 1 | |
| 08-02-95 | Point Loma | 7 | C | | ELMU | ? | A | 39.5 | 132 | 0004 | N | SB-SW | R | Y | 1 | |
| 08-02-95 | Point Loma | 14 | 1A | | UTST | ? | A | 8.5 | 48 | 0003 | N | SB-SW | R | Y | 1 | |
| 08-03-95 | Point Loma | 3 | 2A | | UTST | ? | J | 0.6 | 22 | 0004 | N | SB-SW | R | Y | 1 | |
| 08-03-95 | Point Loma | 7 | C | | SCOC | M | A | 10.5 | 71 | 0013 | N | SB-SW | R | Y | 1 | |
| 08-03-95 | Point Loma | 12 | | 1 | ELMU | ? | A | 25.0 | 114 | 0005 | N | SB-SW | R | Y | 1 | |

**Table 5.** Small mammal data spreadsheet.

[Small mammal data are transcribed from a paper data sheet and stored in a Microsoft Excel spreadsheet. See the appropriate section within appendix 2 for the abbreviations or codes and their explanations. g, gram; mm, millimeter; NOCR, *Notiosorex crawfordi*; PECA, *Peromyscus californicus*; MICA, *Microtus californicus*; AB, Anita Burkett; R, released; D, dead; ?, unknown; A, adult; J, juvenile; N, no; Y, yes]

| Date | Site name | Array No. | Bucket | Snake trap | Species | Sex | Age | Weight (g) | Length (mm) | Toe-clip No. | Recap (?) | Collector's initials | Dispo-sition | Tissue (?) | Sample No. | Notes |
|---|---|---|---|---|---|---|---|---|---|---|---|---|---|---|---|---|
| 06/02/98 | Point Loma | 3 | 1A | | NOCR | ? | ? | | | | ? | AB | R | N | 16 | |
| 06/02/98 | Point Loma | 9 | C | | NOCR | ? | ? | | | | ? | AB | R | N | 16 | |
| 06/02/98 | Point Loma | 12 | C | | PECA | ? | ? | | | | ? | AB | R | N | 16 | |
| 06/02/98 | Point Loma | 12 | 3B | | PECA | ? | ? | | | | ? | AB | R | N | 16 | |
| 06/03/98 | Point Loma | 14 | 1A | | NOCR | ? | ? | | | | ? | AB | D | Y | 16 | Eaten |
| 06/03/98 | Point Loma | 14 | 1A | | NOCR | ? | ? | | | | ? | AB | R | N | 16 | |
| 06/03/98 | Point Loma | 10 | 1A | | NOCR | ? | ? | | | | ? | AB | R | N | 16 | |
| 06/03/98 | Point Loma | 12 | 2B | | PECA | ? | A | | | | ? | AB | R | N | 16 | |
| 06/04/98 | Point Loma | 10 | 1A | | NOCR | ? | ? | | | | ? | AB | D | Y | 16 | |
| 06/04/98 | Point Loma | 14 | 1A | | NOCR | ? | ? | | | | ? | AB | R | N | 16 | |
| 06/04/98 | Point Loma | 1 | 3A | | PECA | ? | ? | | | | ? | AB | R | N | 16 | |
| 06/04/98 | Point Loma | 15 | 1A | | NOCR | ? | ? | | | | ? | AB | R | N | 16 | |
| 06/04/98 | Point Loma | 5 | 3A | | NOCR | ? | ? | | | | ? | AB | R | N | 16 | |
| 06/04/98 | Point Loma | 7 | 2A | | NOCR | ? | ? | | | | ? | AB | R | N | 16 | |
| 06/04/98 | Point Loma | 10 | C | | NOCR | ? | ? | | | | ? | AB | R | N | 16 | |
| 06/04/98 | Point Loma | 10 | 2B | | MICA | ? | ? | | | | ? | AB | R | N | 16 | |
| 06/04/98 | Point Loma | 9 | C | | NOCR | ? | ? | | | | ? | AB | R | N | 16 | |
| 06/04/98 | Point Loma | 8 | | 3 | PECA | ? | A | | | | ? | AB | R | N | 16 | |
| 06/05/98 | Point Loma | 9 | C | | NOCR | ? | ? | | | | ? | AB | R | N | 16 | |
| 06/05/98 | Point Loma | 12 | C | | PECA | ? | A | | | | ? | AB | R | N | 16 | |
| 06/06/98 | Point Loma | 7 | 1A | | NOCR | ? | ? | | | | ? | AB | R | N | 16 | |
| 06/06/98 | Point Loma | 7 | 1B | | NOCR | ? | ? | | | | ? | AB | R | N | 16 | |
| 06/06/98 | Point Loma | 12 | 3B | | PECA | ? | J | | | | ? | AB | R | N | 16 | |
| 06/06/98 | Point Loma | 12 | 3B | | PECA | ? | J | | | | ? | AB | R | N | 16 | |
| 06/06/98 | Point Loma | 12 | C | | PECA | ? | ? | | | | ? | AB | R | N | 16 | |
| 06/06/98 | Point Loma | 12 | 3A | | NOCR | ? | ? | | | | ? | AB | R | N | 16 | |
| 06/06/98 | Point Loma | 10 | 1B | | NOCR | ? | ? | | | | ? | AB | R | N | 16 | |

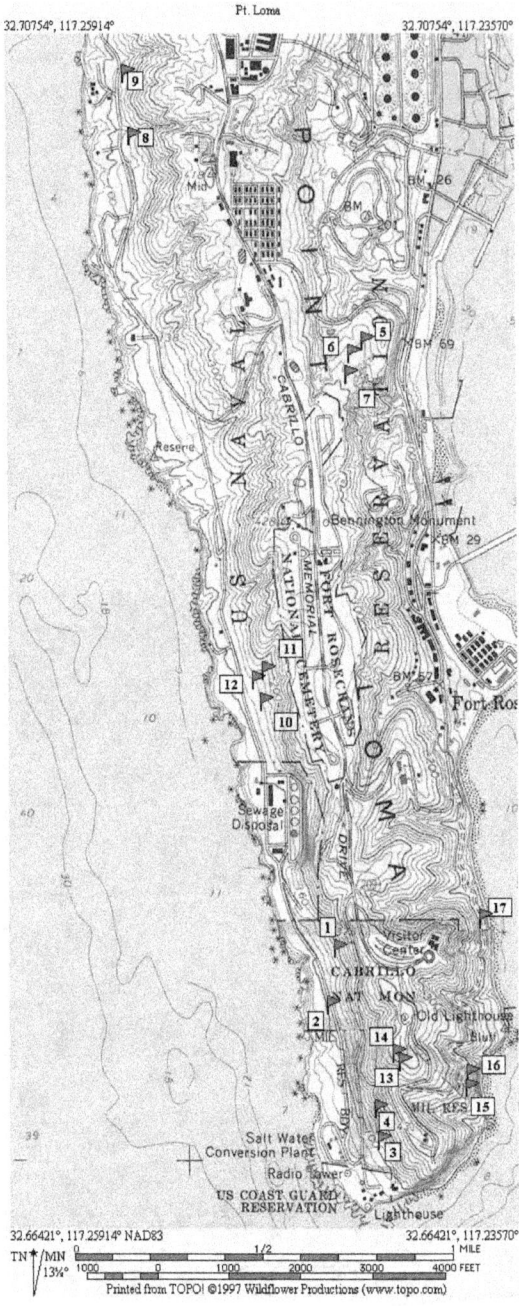

**Figure 22.**   Example site map.

**Table 6.**   Site location file.

[For each study site, a location file is generated, containing information on the sampling history of the site, array locations, and a brief account of the study site and the parties responsible. If the equipment in the field is sensitive or relates to sensitive species, the accuracy of array locations, latitude and longitude, may need to be reduced to protect the integrity of the study. In cases where the data will be available to the public, asterisks may be used to show the precision of the coordinate positions have been reduced to protect the equipment in the field.]

| Site name: | Point Loma Ecological Reserve |
| --- | --- |
| County: | San Diego |
| Responsible parties: | National Park Service |
| Contact: | ###### ##### |
| Phone: | (###) ###-#### |
| Description: | Elevation 22–113 meters. This site is a medium size fragment with arrays sampling 116 hectares and is isolated by urban areas from any other natural lands. The majority of the habitat present is maritime succulent scrub and coastal sage scrub. Some chamise chaparral, grassland and a wash are present also. The Point Loma research area is internally subdivided by roads and buildings, and has a lot of public activity. Point Loma is part of a joint federal partners planning area. |

**Start dates for Sample Periods:**

| | | | | | | |
| --- | --- | --- | --- | --- | --- | --- |
| 08-01-95 | 02-20-96 | 02-18-97 | 02-03-98 | 01-27-99 | 02-09-00 | 02-21-01 |
| 09-12-95 | 04-23-96 | 04-22-97 | 03-31-98 | 04-06-99 | 04-05-00 | 06-21-01 |
| 11-13-95 | 06-25-96 | 06-24-97 | 06-02-98 | 06-08-99 | 06-06-00 | 07-30-01 |
| | 08-27-96 | 09-03-97 | 08-04-98 | 08-18-99 | 08-29-00 | 08-28-01 |
| | 10-29-96 | 10-28-97 | 10-06-98 | 11-10-99 | 11-10-00 | 10-23-01 |
| | | | | | | 11-27-01 |
| | | | | | | 12-18-01 |

**Number of Sample Days:**     320

**Location:**

| Array | Latitude (N) | Longitude (W) | Latitude (N) dec. | Longitude (W) dec. | Elevation (meter) | Datum |
| --- | --- | --- | --- | --- | --- | --- |
| 1 | 32°40'**.** | 117°14'**.** | 32.683** | 117.247** | 26 | NAD 83 |
| 2 | 32°40'**.** | 117°14'**.** | 32.672** | 117.244** | 22 | NAD 83 |
| 3 | 32°40'**.** | 117°14'**.** | 32.667** | 117.242** | 22 | NAD 83 |
| 4 | 32°40'**.** | 117°14'**.** | 32.668** | 117.242** | 33 | NAD 83 |
| 5 | 32°41'**.** | 117°14'**.** | 32.695** | 117.242** | 83 | NAD 83 |
| 6 | 32°41'**.** | 117°14'**.** | 32.695** | 117.243** | 82 | NAD 83 |
| 7 | 32°41'**.** | 117°14'**.** | 32.694** | 117.243** | 90 | NAD 83 |
| 8 | 32°42'**.** | 117°15'**.** | 32.703** | 117.253** | 33 | NAD 83 |
| 9 | 32°40'**.** | 117°14'**.** | 32.674** | 117.244** | 42 | NAD 83 |
| 10 | 32°42'**.** | 117°15'**.** | 32.705** | 117.253** | 35 | NAD 83 |
| 11 | 32°41'**.** | 117°14'**.** | 32.684** | 117.247** | 37 | NAD 83 |
| 12 | 32°41'**.** | 117°14'**.** | 32.683** | 117.247** | 31 | NAD 83 |
| 13 | 32°40'**.** | 117°14'**.** | 32.669** | 117.241** | 109 | NAD 83 |
| 14 | 32°40'**.** | 117°14'**.** | 32.670** | 117.241** | 113 | NAD 83 |
| 15 | 32°40'**.** | 117°14'**.** | 32.668** | 117.238** | 54 | NAD 83 |
| 16 | 32°40'**.** | 117°14'**.** | 32.669** | 117.238** | 55 | NAD 83 |
| 17 | 32°40'**.** | 117°14'**.** | 32.675** | 117.237** | 37 | NAD 83 |

**Table 7.**  Summary of small mammal species captured at each pitfall array.

[Small mammal captures can be viewed as the number of times each species was detected at each array.]

| Species | Array No. | | | | | | | | | | | | | | | | | Total |
|---|---|---|---|---|---|---|---|---|---|---|---|---|---|---|---|---|---|---|
| | 1 | 2 | 3 | 4 | 5 | 6 | 7 | 8 | 9 | 10 | 11 | 12 | 13 | 14 | 15 | 16 | 17 | |
| Woodrat *Neotoma species* | | | | | | | | | | | | 1 | | | | | | 1 |
| Desert woodrat *Neotoma lepida* | | 1 | | | | | | | | | | | | | | | | 1 |
| Deer mouse *Peromyscus species* | | 3 | 5 | 6 | 1 | | 4 | 9 | 4 | 4 | 4 | 4 | | | 6 | 1 | 1 | 52 |
| Deer mouse *Peromyscus maniculatus* | | | | 2 | 1 | | | 1 | | 3 | | | 1 | | 3 | 1 | | 12 |
| California mouse *Peromyscus californicus* | 6 | 3 | 4 | 12 | 9 | 4 | 6 | 8 | 20 | 3 | 3 | 15 | 12 | 3 | 3 | 5 | 6 | 122 |
| Cactus mouse *Peromyscus eremicus* | 4 | 7 | | 1 | 2 | 1 | 1 | 7 | 4 | 1 | 2 | 2 | 3 | 1 | | | 3 | 39 |
| Desert shrew *Notiosorex crawfordi* | 3 | 13 | 7 | 10 | 15 | 1 | 32 | 32 | 25 | 29 | 15 | 25 | 8 | 34 | 11 | 9 | 12 | 281 |
| Ornate shrew *Sorex ornatus* | 1 | | 2 | | | | | 1 | | 2 | | 2 | 1 | 2 | 1 | 1 | 1 | 14 |
| Unknown shrew | | 1 | | | 2 | | | 2 | | | | | | 1 | 2 | | | 13 |
| California vole *Microtus californicus* | 1 | 1 | 1 | 2 | | 2 | 2 | | 4 | 10 | 6 | 9 | | 1 | 1 | 1 | 1 | 42 |
| Western harvest mouse *Reithrodontomys megalotis* | 4 | 2 | 4 | 2 | 23 | 5 | 18 | 15 | 15 | 25 | 17 | 12 | 2 | 5 | 3 | 1 | | 153 |
| Kangaroo rat *Dipodomys species* | | | | | | | | | | | | | | | 1 | | 1 | 2 |
| Desert cottontail *Sylvilagus audubonii* | | | | | | | 1 | | | | 1 | | | | | | | 2 |
| Pocket mouse *Chaetodipus species* | | | | | 1 | 2 | 5 | | | 2 | | | | | | 1 | 1 | 12 |
| California pocket mouse *Chaetodipus californicus* | | | | 2 | | | 3 | | | 1 | | | 1 | | | 1 | | 8 |
| San Diego pocket mouse *Chaetodipus fallax* | | | | | 1 | 1 | | | | | | | | | | | | 2 |
| Little pocket mouse *Perognathus longimembris* | | | | | | | | | | 2 | | | | | | | | 2 |
| House mouse *Mus musculus* | 1 | | 1 | 1 | | | | | | | | | | | | | | 3 |
| Unknown mammal | 2 | | | | | | | | | | | | | | | | | 2 |
| Unknown mouse | | | | | | | 1 | | | | 1 | | | 1 | | | 3 | 6 |
| Unknown rodent | | 3 | | 2 | 2 | | | 1 | 4 | 6 | 1 | | | | 1 | | 2 | 54 |
| Number of individuals | 22 | 34 | 25 | 40 | 58 | 16 | 73 | 77 | 76 | 88 | 50 | 70 | 29 | 48 | 32 | 21 | 31 | 827 |

**Table 8.**  Summary of reptile and amphibian species captured at each pitfall array.

[Reptile and amphibian captures can be viewed as the number of times each species was detected at each array.]

| Species | 1 | 2 | 3 | 4 | 5 | 6 | 7 | 8 | 9 | 10 | 11 | 12 | 13 | 14 | 15 | 16 | 17 | Total |
|---|---|---|---|---|---|---|---|---|---|---|---|---|---|---|---|---|---|---|
| Garden slender salamander *Batrachoseps major* | | 5 | | | | 2 | 8 | 5 | 36 | 2 | 19 | 4 | 1 | 3 | | 1 | 3 | 89 |
| California legless lizard *Anniella pulchra* | | | 2 | | | | 1 | | | | | | | | | | | 3 |
| Southern alligator lizard *Elgaria multicarinata* | 10 | 16 | 25 | 17 | 2 | | 6 | 4 | 7 | 7 | 13 | 16 | 22 | 8 | 11 | 7 | 3 | 174 |
| Western skink *Eumeces skiltonianus* | | | | | | | | | | | | | | | | | | |
| Orange-throated whiptail *Cnemidophorus hyperythrus* | 65 | 28 | 153 | 46 | 1 | 1 | | 2 | 19 | 7 | 3 | 1 | 29 | 55 | 24 | 27 | 11 | 472 |
| Western fence lizard *Sceloporus occidentalis* | 18 | 13 | 54 | 48 | 30 | 18 | 33 | 25 | 28 | 14 | 23 | 18 | 53 | 65 | 40 | 36 | 36 | 552 |
| Side-blotched lizard *Uta stanburiana* | 6 | 42 | 39 | 22 | 52 | 80 | 13 | 4 | 6 | 6 | 2 | 3 | 19 | 49 | 43 | 25 | 23 | 434 |
| Coast horned lizard *Phrynosoma coronatum* | | | | | | | | | | | | | | | | | | |
| California glossy snake *Arizona elegans* | | | | | | | | | | | | | | | | | | |
| Western yellow-bellied racer *Coluber constrictor* | | | | | | | | | | | | | | | | | | |
| Western ringneck snake *Diadophis punctatus* | | 1 | | | | | | 1 | 2 | | 2 | | | | | | 1 | 7 |
| Night snake *Hypsiglena torquata* | | | | | | | | | | 1 | | | | | | | | 1 |
| California kingsnake *Lampropeltis getula* | | 1 | 2 | 1 | | | | | | | | | 1 | | | | | 5 |
| Coachwhip/red racer *Masticophis flagellum* | | | | | | | | | | | | | | | | | | |
| Striped racer *Masticophis lateralis* | 3 | 1 | 3 | 1 | 4 | | 3 | 4 | 7 | 5 | 1 | 1 | 3 | 3 | 3 | | 1 | 43 |
| San Diego gopher snake *Pituophis catenifer* | 1 | 1 | 1 | | | | | 1 | 1 | | | | | | | 1 | | 6 |
| Long-nosed snake *Rhinocheilus lecontei* | | | | | | | | | | | | | | | | | | |
| Red diamond rattlesnake *Crotalus ruber* | | | | | | | | | | | | | | | | | | |
| Southern Pacific rattlesnake *Crotalus viridis* | | 1 | | | 2 | | | | | 1 | | | 1 | | 1 | | | 6 |
| Total Individuals | 103 | 109 | 280 | 135 | 91 | 101 | 64 | 46 | 106 | 44 | 64 | 43 | 129 | 183 | 122 | 97 | 78 | 1,795 |
| Total Species | 6 | 10 | 8 | 6 | 6 | 4 | 6 | 8 | 8 | 8 | 7 | 6 | 8 | 6 | 6 | 6 | 7 | 12 |

This produces a daily average capture rate for each species at each array and across the site as a whole (table 9). In this example, the data are averaged for the entire site. For a comparison of capture rates between sites, one must consider the number of arrays at the sites being compared. For example, the Point Loma site has 17 study arrays, Mission Trails Regional Park has 5 arrays, and the San Diego Wild Animal Park has 20 arrays. To correct for this difference, the total capture rate per species per day is divided by the number of arrays at the study site. Table 10 shows the comparison of the three sites noted above (Point Loma, Mission Trails Regional Park, and San Diego Wild Animal Park).

## 7.4 Data Analysis

Data generated from the described survey methods can be used to address a variety of conservation or management questions. Because the pitfall array method is a passive method of sampling smaller terrestrial vertebrates, the bias is eliminated from observer variability. This is particularly beneficial for studies over large landscapes and timescales, where many people are involved in field sampling.

Data can be used to learn important demographic and life history information for species, such as reproductive cycles, age class characteristics, seasonal activity patterns, survivorship, species interactions, and habitat associations (Fisher and others, 2002; Williams and others, 2002). Species-specific mark-recapture models can be assessed to understand population dynamics and demography and to test specific ecological hypotheses (White and Burnham, 1999; Burnham and Anderson, 2002; and Amstrup and others, 2005). Likewise, single and multi-state occupancy models can be used to understand species metapopulation and patch occupancy dynamics (MacKenzie and others, 2006).

Multivariate statistics can be used to model small terrestrial vertebrate community composition and biodiversity patterns (Clarke and Warwick, 2001; McCune and Grace, 2002; Clarke and others, 2006). Dodd and others (2007) used these types of methods to show changes in the reptile and amphibian communities of northwest Florida over a 28-year period.

Using these and other types of analyses, scientifically robust sampling programs can be designed to define suitable habitat, as well as to understand the effects of landscape characteristics, environmental factors, non-native species, large scale disturbance, human effects, and management actions on wildlife populations. These programs are useful in an adaptive monitoring and management approach. Findings from monitoring studies can lead to specific management recommendations. In turn, the effectiveness of management actions can be tested by continued monitoring, resulting in a science-management feedback loop (Williams and others, 2002).

The pitfall array, as described in this report, is a proven, effective, and unbiased sampling tool that is valuable for a multitude of research projects. Before beginning a project, it is important to define the research objectives, the study design, and the planned analytical treatments. Ecology and statistical texts can give more information on experimental design and statistical analyses of data (for example, Maxwell and Delaney, 2000; Clarke and Warwick, 2001; Williams and others, 2002; MacKenzie and others, 2006).

# 8.0 Acknowledgments

We thank B. Metts, P. Medica, J. Lovich, E. Muths, and A. Backlin for feedback on the earlier versions of this manuscript. Their comments and suggestions greatly improved the final version of this report.

**Table 9.** Capture rate.

[Average capture rate per day can be calculated to show relative abundance of each species at each array and across the study site as a whole.]

| Species | 1 | 2 | 3 | 4 | 5 | 6 | 7 | 8 | 9 | 10 | 11 | 12 | 13 | 14 | 15 | 16 | 17 | Total |
|---|---|---|---|---|---|---|---|---|---|---|---|---|---|---|---|---|---|---|
| **Salamander** | | | | | | | | | | | | | | | | | | |
| Garden slender salamander *Batrachoseps major* | | 0.017 | | | | 0.007 | 0.027 | 0.017 | 0.122 | 0.007 | 0.065 | 0.014 | 0.003 | 0.010 | | 0.003 | 0.010 | 0.018 |
| **Lizard** | | | | | | | | | | | | | | | | | | |
| California legless lizard *Anniella pulchra* | | | 0.007 | | | | 0.003 | | | | | | | | | | | 0.001 |
| Southern alligator lizard *Elgaria multicarinata* | 0.034 | 0.054 | 0.082 | 0.054 | 0.007 | | 0.020 | 0.014 | 0.024 | 0.024 | 0.041 | 0.051 | 0.075 | 0.027 | 0.037 | 0.020 | 0.010 | 0.034 |
| Western skink *Eumeces skiltonianus* | | | | | | | | | | | | | | | | | | |
| Orange-throated whiptail *Cnemidophorus hyperythrus* | 0.218 | 0.095 | 0.510 | 0.153 | 0.003 | 0.003 | | 0.007 | 0.061 | 0.024 | 0.010 | 0.003 | 0.099 | 0.184 | 0.082 | 0.088 | 0.037 | 0.093 |
| Western fence lizard *Sceloporus occidentalis* | 0.058 | 0.044 | 0.167 | 0.143 | 0.095 | 0.054 | 0.105 | 0.082 | 0.095 | 0.044 | 0.078 | 0.054 | 0.177 | 0.214 | 0.126 | 0.119 | 0.122 | 0.105 |
| Side-blotched lizard *Uta stansburiana* | 0.017 | 0.129 | 0.119 | 0.065 | 0.173 | 0.255 | 0.037 | 0.010 | 0.007 | 0.017 | 0.007 | 0.007 | 0.058 | 0.153 | 0.129 | 0.075 | 0.075 | 0.078 |
| Coast horned lizard *Phrynosoma coronatum* | | | | | | | | | | | | | | | | | | |
| **Snake** | | | | | | | | | | | | | | | | | | |
| California glossy snake *Arizona elegans* | | | | | | | | | | | | | | | | | | |
| Western yellow-bellied racer *Coluber constrictor* | | | | | | | | | | | | | | | | | | |
| Western ringneck snake *Diadophis punctatus* | | | | | | | | 0.003 | 0.007 | | 0.007 | | | | | | 0.003 | 0.001 |
| Night snake *Hypsiglena torquata* | | | | | | | | | | 0.003 | | | | | | | | 0.000 |
| California kingsnake *Lampropeltis getula* | | | | | | | | | | | | | 0.003 | | | | | 0.001 |
| Coachwhip/red racer *Masticophis flagellum* | | | | | | | | | | | | | | | | | | |
| Striped racer *Masticophis lateralis* | 0.010 | 0.003 | 0.010 | 0.003 | 0.014 | | 0.010 | 0.014 | 0.024 | 0.017 | 0.003 | 0.003 | 0.010 | 0.010 | 0.010 | | 0.003 | 0.009 |
| San Diego gopher snake *Pituophis catenifer* | 0.003 | 0.003 | 0.003 | | | | | 0.003 | 0.003 | | | | | | | 0.003 | | 0.001 |
| Long-nosed snake *Rhinocheilus lecontei* | | | | | | | | | | | | | | | | | | |
| Red diamond rattlesnake *Crotalus ruber* | | | | | | | | | | | | | | | | | | |
| Southern Pacific rattlesnake *Crotalus viridis* | | | | | 0.007 | | | | | 0.003 | | | 0.003 | | 0.003 | | | 0.001 |

**Table 10.**   Average capture rate table.

[A comparison of average capture rates (per array per day) for 19 species among three study sites. Standard errors could be included for statistical comparison.]

| Species | Average capture rate per array per day | | |
|---|---|---|---|
| | Point Loma | San Diego Wild Animal Park | Mission Trails Regional Park |
| **Salamander** | | | |
| Garden slender salamander | 0.0178 | 0.0052 | |
| *Batrachoseps major* | | | |
| **Lizard** | | | |
| California legless lizard | 0.0006 | 0.0018 | |
| *Anniella pulchra* | | | |
| Southern alligator lizard | 0.0338 | 0.0084 | 0.0324 |
| *Elgaria multicarinata* | | | |
| Western skink | | 0.0332 | 0.0324 |
| *Eumeces skiltonianus* | | | |
| Orange-throated whiptail | 0.0928 | 0.5100 | 0.1853 |
| *Cnemidophorus hyperythrus* | | | |
| Western fence lizard | 0.1046 | 0.0703 | 0.1118 |
| *Sceloporus occidentalis* | | | |
| Side-blotched lizard | 0.0784 | 0.0474 | 0.0647 |
| *Uta stanburiana* | | | |
| Coast horned lizard | | 0.0215 | |
| *Phrynosoma coronatum* | | | |
| **Snake** | | | |
| California glossy snake | | | |
| *Arizona elegans* | | | |
| Western yellow-bellied racer | | | |
| *Coluber constrictor* | | | |
| Western ringneck snake | 0.0012 | 0.0011 | |
| *Diadophis punctatus* | | | |
| Night snake | 0.0002 | 0.0002 | |
| *Hypsiglena torquata* | | | |
| California kingsnake | 0.0010 | 0.0065 | |
| *Lampropeltis getula* | | | |
| Coachwhip/Red racer | | 0.0008 | |
| *Masticophis flagellum* | | | |
| Striped racer | 0.0086 | 0.0189 | 0.0059 |
| *Masticophis lateralis* | | | |
| San Diego gopher snake | 0.0012 | 0.0045 | |
| *Pituophis catenifer* | | | |
| Long-nosed snake | | 0.0019 | |
| *Rhinocheilus lecontei* | | | |
| Red diamond rattlesnake | | 0.0048 | 0.0029 |
| *Crotalus ruber* | | | |
| Southern Pacific rattlesnake | 0.0010 | 0.0013 | |
| *Crotalus viridis* | | | |

# 9.0 References Cited

Amstrup, S.C., McDonald, T.L., and Manly, B.F., 2005, Handbook of capture-recapture analysis: Princeton, New Jersey, Princeton University Press, 313 p.

Banta, B.H., 1957, A simple trap for collecting desert reptiles: Herpetologica, v. 13, p. 174-176.

Banta, B.H., 1962, A preliminary account of the herpetofauna of the Saline Valley hydrographic basin, Inyo County, California: The Wasmann Journal of Biology, v. 20, p. 161-251.

Bennett, D.H., Gibbons, J.W., and Franson, J.C., 1970, Terrestrial activity of aquatic turtles: Ecology, v. 51, no. 4, p. 738-740.

Bostic, D.L., 1965, Home range of the Teiid lizard, *Cnemidophorus hyperythrus beldingi*: The Southwestern Naturalist, v. 10, no. 4, p. 278-281.

Burke, V.J., Rathbun, S.L.. Bodie, J.R., and Gibbons, J.W., 1998, Effect of density on predation rate for turtle nests in a complex landscape: Oikos, v. 83, p. 3-11.

Bostic, D.L., 1965, Home range of the Teiid lizard, *Cnemidophorus hyperythrus beldingi*: The Southwestern Naturalist, v. 10, no. 4, p. 278-281.

Burkett, D.W., and Thompson, B.C., 1994, Wildlife association with human-altered water sources in semiarid vegetation communities: Conservation Biology, v. 8, p. 682-690.

Burnham, K.P., and Anderson, D.R., 2002, Model selection and multimodel inference. A Practical Information-Theoretical Approach (2nd ed.): Springer-Verlag, New York, 448 p.

Bury, R.B., and Corn, P.S., 1987, Evaluation of pitfall trapping in northwestern forests: Trap arrays with drift fences: Journal of Wildlife Management, v. 51, p. 112-119.

Cagle, F.R., 1939, A system for marking turtles for future identification: Copeia, v. 1939, p. 170-173.

Campbell, H.W. and Christman, S.P., 1982, Field techniques for herpetofaunal community analysis, *in* Scott, N.J., Jr., ed., Herpetological communities: U.S. Fish and Wildlife Service Wildlife Research Report 13, p. 193-200.

Case, T.J. and Fisher, R.N., 2001, Measuring and predicting species presence: Coastal sage scrub case study, *in* Hunsaker, C.T., Goodchild, M.F., Friedl, M.A. and Case, T.J., eds., Spatial Uncertainty in Ecology: Springer-Verlag, New York, p. 47-71.

Clarke, K.R., Somerfield, P.J., and Chapman, M.G., 2006, On resemblance measures for ecological studies, including taxonomic dissimilarities and a zero-adjusted Bray-Curtis coefficient for denuded assemblages: Journal of Experimental Marine Biology and Ecology, v. 330, p. 55-80.

Clarke, K.R., and Warwick, R.M., 2001, Change in marine communities. An approach to statistical analysis and interpretation, (2nd ed): PRIMER-E, Plymouth, UK, 172 p.

Corn, P.S., and Bury, R.B., 1990, Sampling methods for terrestrial amphibians and reptiles: U.S. Department of Agriculture Forest Service General Technical Report PNW-GTR-256, 34 p.

Corn, P.S., 1994, Straight-line drift fences and pitfall traps, *in* Heyer, W.R., Donnelly, M.A., McDiarmid, R.W., Hayek, L.C., and Foster, M.S., eds., Measuring and monitoring biological diversity—Standard Methods for Amphibians: Smithsonian Institution Press, Washington, D.C., p. 109-117.

Crawford, E., and Kurta, A., 2000, Color of pitfall affects trapping success for anurans and shrews: Herpetological Review, v. 31, no. 4, p. 222-224.

DeGraaf, R.M., and Rudis, D.D., 1990, Herpetofaunal species composition and relative abundance among three New England forest types: Forest Ecology and Management v. 32, p. 155-165.

Dodd, C.K., Jr., 1992, Biological diversity of a temporary pond herpetofauna in north Florida sandhills: Biodiversity and Conservation v. 1, p. 125-142.

Dodd, C.K., Jr., 1994, Monitoring and protecting biotic diversity, *in* Majumdar, S.K., Brenner, F.J., Lovich, J.E., Schalles, J.F., and Miller, E.W., eds., Biological diversity—Problems and challenges: Pennsylvania Academy of Science, Easton, Pennsylvania, p. 1-11.

Dodd, C.K., Jr., Barichivich, W.J., Johnson, S.A., and Staiger, J.S., 2007, Changes in a northwestern Florida Gulf Coast herpetofaunal community over a 28-y period: American Midland Naturalist, v. 158, p. 29-48.

Dodd, C.K., Jr., and Scott, D.E., 1994, Drift fences encircling breeding sites, *in* Heyer, W.R., Donnelly, M.A., McDiarmid, R.W., Hayek, L.C., and Foster, M.S., eds., Measuring and monitoring biological diversity—Standard methods for amphibians: Smithsonian Institution Press, Washington, D.C., p. 125-130.

Douglas, M.E., 1979, Migration and sexual selection in *Ambystoma jeffersonianum*: Canandian Journal of Zoology, v. 57, no. 12, p. 2303-2310.

Enge, K.M., 1997, Use of silt fencing and funnel traps for drift fences: Herpetological Review, v. 28, no. 1, p. 30-31.

Enge, K.M., 2001, The pitfalls of pitfall traps: Journal of Herpetology, v. 35, p. 467-478.

Fair, W.S., and Henke, S.E., 1997, Efficacy of capture methods for a low density population of *Phrynosoma cornutum*: Herpetological Review, v. 28, no. 3, p. 135-137.

Fellers, G.M., and Pratt, D., 2002, Terrestrial vertebrate inventory, Point Reyes National Seashore, 1998-2001: National Park Service Report. 73 p.

Fisher, R.N., and Case, T.J., 1997, A field guide to the reptiles and amphibians of coastal southern California: Lazer Touch, San Mateo, California. 46 p.

Fisher, R.N., and Case, T. J., 2000, Distribution of the herpetofauna of coastal southern California with reference to elevation effects, *in* Keeley, J.E., Baer-Keeley, M., and Fotheringham, C.J., eds., Second interface between ecology and land development in California: U.S. Geological Survey Open-File Report 00-62, p. 137-143.

Fisher, R.N., and Shaffer, H.B., 1996, The decline of amphibians in California's great central valley: Conservation Biology, v. 10, p. 1387-1397.

Fisher, R.N., Suarez, A.V., and Case, T.J., 2002, Spatial patterns in the abundance of the coastal horned lizard: Conservation Biology, v.16, p. 205-215.

Fitch, H.S., 1951, A simplified type of funnel trap for reptiles: Herpetologica, v. 7, p. 77-80.

Friend, G.R., 1984, Relative abundance of two pitfall-drift fence systems for sampling small vertebrates: Australian Zoologist, v. 21, no. 5, p. 423-434.

Gibbons, J.W., 1970, Terrestrial activity and the population dynamics of aquatic turtles: American Midland Naturalist, v. 83, no. 2, p. 404-414.

Gibbons, J.W. and Bennett, D.H., 1974, Determination of anuran terrestrial activity patterns by a drift fence method: Copeia, v. 1, p. 236-243.

Gibbons, J.W., and Semlitsch, R.D., 1981, Terrestrial drift fences with pitfall traps—An effective technique for quantitative sampling of animal populations: Brimleyana, v. 7, p. 1-16.

Gill, D.E., 1978, The metapopulation ecology of the red-spotted newt, *Notophthalmus viridescens* (Rafinesque): Ecological Monographs, v. 48, no. 2, p. 145-166.

Gloyd, H.K., 1947, Some rattlesnake dens of South Dakota: Chicago Naturalist v. 9, p. 87-97.

Heyer, W.R., Donnelly, M.A., McDiarmid, R.W., Hayek, L.C., and Foster, M.S., 1994, Measuring and monitoring biological diversity—Standard methods for amphibians: Smithsonian Institution Press, Washington, D.C., 364 p.

Imler, R.I., 1945, Bullsnakes and their control on a Nebraska wildlife refuge: Journal of Wildlife Management v. 9, no. 4, p. 265-273.

Ireland, T.T., Wolters, G.L., and Schemnitz, S.D., 1994, Recolonization of wildlife on a coal strip-mine in northwestern New Mexico: Southwestern Naturalist, v. 39, p. 53-57.

Jockusch, E.L., and Wake, D.B., 2002, Falling apart and merging—Diversification of slender salamanders (Plethodontidae—Batrachoseps) in the American West: Biological Journal of the Linnean Society, v. 76, p. 361-391.

Jones, K.B., 1981, Effects of grazing on lizard abundance and diversity in western Arizona: Southwestern Naturalist, v. 26, p. 107-115.

Jorgensen, E.E., Vogel, M., and Demarais, S. A., 1998, A comparison of trap effectiveness for reptile sampling: Texas Journal of Science, v. 50, no. 3, p. 235-242.

Laakkonen, J., Fisher, R.N., and Case, T.J., 2001, Effect of land cover, habitat fragmentation, and ant colonies on the distribution and abundance of shrews in southern California: Journal of Animal Ecology, v. 70, p. 776-788.

Loredo, I., Vuren, D.V., and Morrison, M. L., 1996, Habitat use and migration behavior of the California tiger salamander: Journal of Herpetology, v. 30, no. 2, p. 285-288.

MacKenzie, D.I., Nichols, J.D., Royle, J.A., Pollock, K.H., Bailey, L.L., and Hines, J.E., 2006, Occupancy estimation and modeling—Inferring patterns and dynamics of species occurrence: Elsevier, San Diego, California, 324 p.

Mahoney, M.J., Parks, D.S.M., and Fellers, G.M., 2003, *Uta stansburiana* and *Elgaria multicarinata* on the California Channel Islands—Natural dispersal or artificial introduction: Journal of Herpetology, v. 37, p. 586-591.

Maldonado, J. E., Vila, C., and Wayne, R.K., 2001, Tripartite genetic subdivisions in the ornate shrew (*Sorex ornatus*): Molecular Ecology, v. 10, p. 127-147.

Maxwell, S.E., and Delaney, H.D., 2000, Designing experiments and analyzing data: Lawrence Erlbaum Associates, Inc., Mahwah, New Jersey, 902 p.

McCoy, E.D. and Mushinsky, H.R., 1994, Effects of fragmentation on the richness of vertebrates in the Florida scrub habitat: Ecology, v. 75, p. 446-457.

McCune, B., and Grace, J.B., 2002, Analysis of ecological communities: MJM Press, Gleneden Beach, Oregon, 300 p.

Medica, P.A., Hoddenbach, G.A., and Lannom Jr., J.R., 1971, Lizard sampling techniques: Rock Valley Miscellaneous Publications no. 1, 55 p.

Milstead, W.W., 1953, Ecological distributions of the lizards of the La Mota Mountain region of Trans-Pecos Texas: Texas Journal of Science, v. 5, p. 403-415.

Murphy, C.G., 1993, A modified drift fence for capturing treefrogs: Herpetological Review, v. 24, no. 4, p. 143-145.

Nelson, D.H., and Gibbons, J.W., 1972, Ecology, abundance, and seasonal activity of the scarlet snake, *Cemophora coccinea*: Copeia, v. 1972, no. 3, p. 582-584.

Parker, W.S., 1972, Aspects of the ecology of a sonoran desert population of the western banded gecko, *Coleonyx variegatus* (Sauria, Eublepharinae): The American Midland Naturalist, v. 88, no. 1, p. 209-224.

Pearson, P.G., 1955, Population ecology of the spadefoot toad, *Scaphiopus h. holbrooki* (Harlan): Ecological Monographs, v. 25, no. 3, p. 233-267.

Richmond, J.Q. and Reeder, T.W., 2002, Evidence for parallel ecological speciation in the scincid lizards of the *Eumeces skiltonianus* species group (Squamata: Scincidae): Evolution, v. 56, p. 1498-1513.

Rice, C.G., Jorgensen, E.E., and Demarais, S., 1994, A comparison of herpetofauna detection and capture techniques in southern New Mexico: Texas Journal of Agriculture and Natural Resources, v. 7, p. 107-113.

Rudolph, D.C., and Dickson, J.G., 1990, Streamside zone width and amphibian and reptile abundance: The Southwest Naturalist, v. 35, p. 472-476.

Ryan, T.J., Philippi, T., Leiden, Y.A., Dorcas, M.E., Wigley, T.B., and Gibbons, J.W., 2002, Monitoring herpetofauna in a managed forest landscape—Effects of habitat types and census techniques: Forest Ecology and Management, v. 167, p. 83-90.

Sawyer, J.O., and Keeler-Wolf, T., 1995, A manual of California vegetation: California Native Plant Society, 471 p.

Scott Jr., N.J., 1982, Herpetological communities—A symposium of the Society for the Study of Amphibians and Reptiles and the Herpetologists' League, August 1977: U.S. Department of Interior Wildlife Research Report 13, 239 p.

Semlitsch, R.D., Brown, K.L., and Caldwell, J.P., 1981, Habitat utilization, seasonal activity, and population size structure of the southeastern crowned snake *Tantilla coronata*: Herpetologica, v. 37, no. 1, p. 40-46.

Shoop, C.R., 1965, Orientation of *Ambystoma maculatum*—movements to and from breeding ponds: Science, v. 149, no. 3683, p. 558-559.

Stebbins, R.C., 1985, A field guide to western reptiles and amphibians: Houghton, Mifflin Co., Boston, Mass., 336 p.

Storm, R.M., and Pimental, R.A., 1954, A method for studying amphibian breeding populations: Herpetologica, v. 10, p. 161-166.

Vogt, R.C., and Hine, R.L., 1982, Evaluation of techniques for assessment of amphibian and reptile populations in Wisconsin, *in* Scott, N.J., Jr. (ed), Herpetological Communities: U.S. Fish and Wildlife Service Wildlife Research Report 13, p. 201-217.

White, G.C. and Burnham, K.P., 1999, Program MARK: Survival estimation from populations of marked animals: Bird Study 46 Supplement, p. 120-138. Also available at http://www.warnercnr.colostate.edu/~gwhite/mark/mark.htm

Williams, B.K., Nichols, J.D., and Conroy, M.J., 2002, Analysis and management of animal populations: Modeling, estimation, and decision making: Academic Press, San Diego, California, 817 p.

Woodbury, A.M., 1951, Symposium: A snake den in Tooele County, Utah—Introduction-a ten year study: Herpetologica, v. 7, no. 1, p. 4-14.

Woodbury, A.M., 1953, Methods of field study in reptiles: Herpetologica, v. 9, no. 2, p. 87-92.

Wygoda, M.L., 1979, Terrestrial activity of striped mud turtles, *Kinosternon baurii* (Reptilia, Testudines, Kinosternidae) in west-central Florida: Journal of Herpetology, v. 13, no. 4, p. 469-480.

Yunger, J.A., Brewer, R., and Snook, R., 1992, A method for decreasing trap mortality of *Sorex*: Canadian Field-Naturalist, v. 106, p. 249-251.

Zug, G.R., Lhon, W.Z., Min, T.Z., Kyaw, K., Thin, T., Win, H., Nyein, M.T.D., Aung, K., and Tin, K.T., 2001, Durability of silt-fencing for drift-fence arrays at a tropical site: Herpetological Review, v. 32, p. 235-236.

**Appendix 1.**   Materials and supplies needed to build, maintain, and sample a pitfall trapping array.

[in, inch; mm, millimeter; cm, centimeter; m, meter; gal, gallon; L, liter; max, maximum; min, minimum; %, percent]

---

### Site and equipment construction supplies

#### Funnel Trap

1.25-in. (32-mm) binder clips

0.25-in. (0.635-cm) mesh hardware cloth, 36-in.    100-ft roll (91.45 cm   30.48 m)

0.25-in. (0.635-cm) mesh hardware cloth,  24-in.    100-ft roll (60.69 cm   30.48 m)

PVC "T" joint, 1.5-in. (3.81-cm) SSS for desert sites

hog ringer pliers

steel hog rings

shake shingles, medium untreated or 0.75-in. (1.91-cm) plywood for desert sites

tin snips

bailing wire

#### Drift Fence

shade cloth, 12-in. (30.48-cm) tall

staples

wooden stakes, 1    2    24 in. (2.54-cm    5.08-cm    60.96-cm)

heavy duty staple gun

scissors

#### Buckets and Lids

2-   2-in. lumber (5.08 cm    5.08 cm)

5-gal (18.9-L) buckets

6-gal (22.7-L) buckets

plastic bucket lids

1.25-in. (3.18-cm) drywall screws

SAE washers #10

bungee cords for desert site

marker to label buckets and lids

#### Tools

gloves

spade shovel

pick axe

sledge hammer

50-m tape measure

0.125-in. (0.318-cm) drill bit

pry bar/digging bar

electric drill

circular saw

Phillips head drill bit

0.5-in. (1.27-cm) drill bit for desert sites

2.25-in. (5.72-cm) drill bit for desert sites

#### Site Operation

cellulose sponges

max-min thermometer or temperature data logger

1.5-in. (3.81-cm) PVC pipe with foam insulation

1-in. (2.54-cm) PVC pipe

#### Field Kit Supplies

notebook

lab marker

carrying case

large Tweezers

small re-closable plastic bags

metric tape measure

straight microscissors

water bottle for sponges

50-ml tissue tubes with 70% ethanol

fiberboard storage box and dividers

1.5-ml tissue tubes with 95% ethanol

extra snake bags or appropriate animal containers

selection of appropriately sized metric spring scales

snake bag

snake hook

small Tweezers

handheld computer

large re-closable plastic bags

appropriate data forms

ice chest with ice packs

clear plastic, metric ruler

fade resistant, water proof pen

appropriate wildlife field guides

small spatula or cup for cleaning out debris from within pitfall traps or bailing out water

#### Weather Station

wood bolt, wood thread on one end and machine threat on the other

wing nut and washer to fit wood bold

2-in. PVC pipe

---

**Appendix 2.**    Species codes and abbreviatoins.

[A list of the species codes and abbreviations used throught the text and their explanations.]

| Category | Code | Scientific name | Common name |
|---|---|---|---|
| **Species** | | | |
| **Reptiles** | | | |
| **Lizards** | | | |
| | ANPU | *Anniella pulchra* | California Legless Lizard |
| | CNHY | *Cnemidophorus hyperythrus* | Orange-Throated Lizard |
| | CNTI | *Cnemidophorus tigris* | Western Whiptail |
| | COVA | *Coleonyx variegatus* | Western Banded Lizard |
| | ELMU | *Elgaria multicarinata* | Southern Alligator Lizard |
| | EUGI | *Eumeces gilberti* | Gilbert's Skink |
| | EUSK | *Eumeces skiltonianus* | Western Skink |
| | GAWI | *Gambelia wislizenii* | Long-Nosed Leopard Lizard |
| | PHCO | *Phrynosoma coronatum* | Coast Horned Lizard |
| | SCOC | *Sceloporus occidentalis* | Western Fence Lizard |
| | SCOR | *Sceloporus orcutti* | Granite Spiny Lizard |
| | UTST | *Uta stansburiana* | Side-Blotched Lizard |
| | XAHE | *Xantusia henshawi* | Granite Night Lizard |
| | XAVI | *Xantusia vigilis* | Desert Night Lizard |
| **Snakes** | | | |
| | AREL | *Arizona elegans* | Glossy Snake |
| | CHBO | *Charina bottae* | Rubber Boa |
| | COCO | *Coluber constrictor* | Racer |
| | CRMI | *Crotalus mitchellii* | Speckled Rattlesnake |
| | CRRU | *Crotalus ruber* | Red Diamond Rattlesnake |
| | CRVI | *Crotalus viridis* | Western Rattlesnake |
| | DIPU | *Diadophis punctatus* | Ringneck Snake |
| | HYTO | *Hypsiglena torquata* | Night Snake |
| | LAGE | *Lampropeltis getula* | Common Kingsnake |
| | LAZO | *Lampropeltis zonata* | California Mountain Kingsnake |
| | LEHU | *Leptotyphlops humilis* | Western Blind Snake |
| | LITR | *Lichanura trivirgata* | Rosy Boa |
| | MAFL | *Masticophis flagellum* | Coachwhip |
| | MALA | *Masticophis lateralis* | California Whipsnake |
| | PIME | *Pituophis catenifer* | Gopher Snake |
| | RHLE | *Rhinocheilus lecontei* | Long-Nosed Snake |
| | SAHE | *Salvadora hexalepis* | Western Patched-Nosed Snake |

| Category | Code | Scientific name | Common name |
|---|---|---|---|
| **Species** | | | |
| **Reptiles—Cont.** | | | |
| **Lizards** | | | |
| | TAPL | *Tantilla planiceps* | California Black-Headed Snake |
| | THEL | *Thamnophis elegans* | Western Terrestrial Garter Snake |
| | THHA | *Thamnophis hammondii* | Two-Striped Garter Snake |
| | THSI | *Thamnophis sirtalis* | Common Garter Snake |
| | TRBI | *Trimorphodon biscutatus* | Lyre Snake |
| **Amphibians** | | | |
| **Frogs** | | | |
| | BUBO | *Bufo boreas* | Western Toad |
| | BUMI | *Bufo californicus* | Southwestern Toad |
| | BUPU | *Bufo punctatus* | Red Spotted Toad |
| | HYCA | *Hyla cadaverina* | California TreeFrog |
| | HYRE | *Hyla regilla* | Pacific Chorus Frog |
| | RAAU | *Rana aurora* | Red-Legged Frog |
| | RACA | *Rana catesbeiana* | Bullfrog |
| | SCHA | *Spea hammondii* | Western Spadefoot |
| **Salamanders** | | | |
| | ANLU | *Aneides lugubris* | Arboreal Salamander |
| | BANI | *Batrachoseps nigriventris* | Black-Bellied Salamander |
| | BAPA | *Batrachoseps pacificus* | Garden Slender Salamander |
| | ENES | *Ensatina eschscholtzii* | Monterey Salamander |
| | TATO | *Taricha torosa* | California Newt |
| **Plants** | | | |
| | ARCA | *Artemisia californica* | California Sagebrush |
| | DUED | *Dudleya edulis* | Ladies' Fingers |
| | DULA | *Dudleya lanceolata* | Coastal/Lanceleaf Dudleya |
| | ENCA | *Encelia californica* | California Encelia |
| | ERFA | *Eriogonum fasciculatum* | California Buckwheat |
| | NNG | – | Unknown Non-native Grass |
| | RHIN | *Rhus integrifolia* | Lemonadeberry |
| **Small Mammals** | | | |
| | MICA | *Microtus californicus* | California Vole |
| | NOCR | *Notiosorex crawfordi* | Grey Shrew |
| | PECA | *Peromyscus californicus* | California Mouse |

| Category | Code | Description |
|---|---|---|
| Sex | ? | Unknown |
| | F | Female |
| | M | Male |
| Age | ? | Unknown |
| | A | Adult |
| | J | Juvenile |
| Recap | ? | Unknown |
| | N | No |
| | Y | Yes |

| Category | Code | Description |
|---|---|---|
| Disposition | ? | Unknown |
| | D | Dead |
| | E | Escaped |
| | R | Released |
| Tissue | ? | Unknown |
| | N | No |
| | Y | Yes |
| Substrate | BR | Bare rock |
| | LL | Leaf litter |
| | OR | Organic soil |
| | SS | Sandy soil |